The 2nd International Conference on Computational Engineering and Intelligent Systems

The 2nd International Conference on Computational Engineering and Intelligent Systems

Editors

Abdelmadjid Recioui
Hamid Bentarzi
Fatma Zohra Dekhandji

MDPI • Basel • Beijing • Wuhan • Barcelona • Belgrade • Manchester • Tokyo • Cluj • Tianjin

Editors
Abdelmadjid Recioui
University M'hamed Bougara of Boumerdes
Boumerdes
Algeria

Hamid Bentarzi
University M'hamed Bougara of Boumerdes
Boumerdes
Algeria

Fatma Zohra Dekhandji
University M'hamed Bougara of Boumerdes
Boumerdes
Algeria

Editorial Office
MDPI
St. Alban-Anlage 66
4052 Basel, Switzerland

This is a reprint of Proceedings published online in the open access journal *Engineering Proceedings* (ISSN 2673-4591) (available at: https://www.mdpi.com/2673-4591/29/1).

For citation purposes, cite each article independently as indicated on the article page online and as indicated below:

LastName, A.A.; LastName, B.B.; LastName, C.C. Article Title. *Journal Name* **Year**, *Volume Number*, Page Range.

ISBN 978-3-0365-8112-5 (Hbk)
ISBN 978-3-0365-8113-2 (PDF)

Contents

About the Editors

Abdelmadjid Recioui

Abdelmadjid Recioui (Professor) is a Professor at the Institute of Electrical Engineering and Electronics University of Boumerdes, Algeria. He obtained a PhD degree in electrical and electronic engineering option telecommunications from the Institute of Electrical Engineering and Electronics, University of Boumerdes in 2011. He holds also a Master's (Magister) degree in Electronic System Engineering which he achieved at the Institute of Electrical Engineering and Electronics, University of Boumerdes in 2006. In June 2002, he finished his engineering studies at the institute of Electrical Engineering and Electronics, University of Boumerdes. He is a research assistant at the laboratory signals and systems from January 2008 to oresent in Laboratory: Signals and Systems, Inst. of Electrical Engineering and electronics, University of Boumerdes. His research interests include: antennas, wireless communication systems, antenna array synthesis and design, capacity enhancement, system optimization, smart antennas, power system protection, power system optimization, power system communications.

Hamid Bentarzi

Hamid Bentarzi (Professor) has been a faculty member in the Institute (IGEE), University M'hamed Bougara, Boumerdes, Algeria, since 1998. He received M. Phil degrees in applied electronics from INELEC, Boumerdes, in 1992 and PhD degree from the ENP, Algiers, in 2004. His research activities are centered upon control and protection systems in the smart grid as well as their reliability enhancement using the recent technologies such as PMU and μPMU. He has been an author and co-author in more than hundred papers and three books.

Fatma Zohra Dekhandji

Fatma Zohra Dekhandji (Dr.) is a faculty member at the institute of Electrical and Electronic Engineering, University M'hamed Bougara of Boumerdes. She obtained her PhD in 2016. Her research interests include: power quality, smart grids, optimization and renewable energies. She authored and co-authored more than 40 research papers and contributed to two books.

engineering proceedings

MDPI

Editorial

The Second International Conference on Computational Engineering and Intelligent Systems (ICCEIS2022) †

Abdelmadjid Recioui *, Hamid Bentarzi and Fatma Zohra Dekhandji

Laboratory of Signals and Systems, Institute of Electrical and Electronic Engineering, University M'hamed Bougara of Boumerdes, Boumerdes 35000, Algeria; hamid.bentarzi@gmail.com (H.B.); fzdekhandji@univ-boumerdes.dz (F.Z.D.)
* Correspondence: a_recioui@univ-boumerdes.dz
† All proceeding papers of this volume are presented at the 2nd International Conference on Computational Engineering and Intelligent Systems, online, 18–20 November 2022.

1. Introduction

Multidisciplinary engineering is currently a proliferating area, whereby focus is placed upon an engineering practice which combines several academic disciplines. Computational engineering is a modern and multidisciplinary science for computer-based modeling, simulation, analysis, and optimization of complex engineering applications and natural phenomena. Regarded as the intersection of computer science and applied math, computational engineering deals with mathematical techniques for the modeling and simulation of complex systems, parallel programming and collaborative software development, and methods for organizing, exploring, visualizing, processing, and analyzing extensive datasets. Computational engineering includes fundamental engineering and science, as well as an advanced knowledge of mathematics, algorithms, and computer languages. On the other hand, intelligent systems engineering offers the next generation of solutions through creating systems that sense and react to their environments via computing and artificial intelligence. Intelligent systems are technologically advanced machines that are designed to respond to some specific requirements.

The aim of this conference is to bring together scientists, researchers, and industrialists to share knowledge and findings with regard to the conference topics. The purpose is to provide a platform for a possible collaboration and exchange of ideas to further advance in this field.

Topics of interest concern, but are not limited to, the following: biomedical engineering and applications; the computational study of biological systems; climate modeling; energy systems; modeling and simulation; multiphysical models and co-simulation; cybersecurity; data science and engineering; high-performance computing; optimization; multiagent systems; evolutionary computation; artificial intelligence; complex systems; computational intelligence and soft computing; intelligent control; advanced control technology; robotics and applications; intelligent information processing; iterative learning control; machine learning; and smart grids and systems.

Citation: Recioui, A.; Bentarzi, H.; Dekhandji, F.Z. The Second International Conference on Computational Engineering and Intelligent Systems (ICCEIS2022). *Eng. Proc.* **2023**, *29*, 19. https://doi.org/10.3390/engproc2023029019

Published: 20 June 2023

2. Conference Digest

Peer-review process and regulations: The review process was single blinded. The main criteria of acceptance were the originality, the clarity of the presentation, and the findings.

The conference took place over two days. The program included one keynote with a question and answer (Q&A) session. A total of 20 papers were scheduled. Eighteen of these papers were presented online.

A plenary presentation was delivered by Prof. Hamid Bentarzi from the Laboratory of Signals and Systems, Institute of Electrical and Electronic Engineering, University M'hamed

Bougara of Boumerdes in Algeria. His contribution entitled "Micro-Synchro-Phasor Measurement Unit: Developments and Trends" stressed that the existing PMU can only be used in transmission grids. However, the μSPMU is also able to store synchrophasor measurements and communicate real-time data at the distribution level of the power grid. This keynote presented two main points:

(1) The development of a μPMU with the precise time-stamping of measurements via GPS to allow the comparison of voltage phase angle, δ (i.e., the precise timing of the voltage waveform), at different locations on the hardware. On the software, the development of a new computational framework to manage large, high-density data streams with nanosecond time-stamping and online capabilities that avoids the need for phasor data concentrators.

(2) The development of many practical applications using this new type of information provided by a μPMU in the distribution network context, including the integration of an intermittent renewable generation such as distributed control, adaptive protection, and the cybersecurity of the electric grid.

Seventeen papers were submitted to the conference and are published in a Special Issue of MDPI's *engineering proceedings* [1–4].

3. Committees

- Honorary chair

  The late Professor Larbi REFOUFI.

- Conference general chair

  Prof. Dr. Abdelmadjid RECIOUI.

- Conference co-chairs:

  Prof. Dr. Hamid BENTARZI;
  Dr. Fatma Zohra DEKHANDJI.

- International scientific committee

  AKSAS R. (Algeria);
  AKKURT I. (Turkey);
  ALBARBAR A. (UK);
  ARSHAD M. K. (Malaysia);
  BELAIDI H. (Algeria);
  BELKHIRI M. (Algeria);
  BELATRECHE A. (UK);
  BENALI K. (Belgium);
  BENAZZOUZ D. (Algeria);
  BENTARZI H. (Algeria);
  BENDIB R. (Algeria);
  BENZID R. (Algeria);
  BOUCHAHDANE M. (Algeria);
  BOULAKROUNE M. (Algeria);
  BOUTEJDAR A. (Germany);
  CHALLAL M. (Algeria);
  CHELLALI F. (Algeria);
  CHOUBANI F. (Tunisia);
  DAAMOUCHE A. (Algeria);
  DAHIMENE A. (Algeria);
  DEKHANDJI F. Z. (Algeria);
  EL MOUSSATI A. (Morocco);
  El- OUALKADI A. (Morocco);
  ESSAIDI M. (Morocco);

FORTAKI T. (Algeria);
GUESSOUM A. (Algeria);
HARICHE K. (Algeria);
HARKAT M. (Algeria);
IKHLEF A. (UK);
KACHA A. (Algeria);
KHEZZAR A. (Algeria);
KIMOUCHE H. (Algeria);
KRIM F. (Algeria);
LACHOURI A. (Algeria);
LIAO K. (USA);
MAACHE A. (Algeria);
MAIDI A. (Algeria);
MAUN J. C. (Belgium);
MEDJOUDJ R. (Algeria);
MEKHALDI A. (Algeria);
MEKHALEF S. (Malaysia);
MELGANI F. (Italy);
METIDJI B. (Algeria);
NAKAMATSU K. (Japan);
OUADI A. (Algeria);
OULD BOUAMMAMA B. (France);
SADOUKI L. (Algeria);
TILMATINE A. (Algeria);
YAGOUB M. C. (Canada);
ZAHAWI B. (UAE);
ZENNIR Y. (Algeria);
ZITOUNI A. (Algeria).

Acknowledgments: We would like to thank all contributors and the Signals and Systems Laboratory for supporting this event.

Conflicts of Interest: The authors declare no conflict of interest.

References

1. ICCEIS2022 Conference Proceedings. Available online: https://www.mdpi.com/2673-4591/29/1 (accessed on 18 June 2023).
2. Dekhandji, F.Z.; Recioui, A.; Ladada, A.; Moulay Brahim, T.S. Detection and Classification of Power Quality Disturbances Using LSTM. *Eng. Proc.* **2023**, *29*, 2. [CrossRef]
3. Recioui, A.; Dekhandji, F.Z. Pricing Policy Applied to Power Quality Enhancement in Smart Metering Systems. *Eng. Proc.* **2023**, *29*, 15. [CrossRef]
4. Touabi, C.; Ouadi, A.; Bentarzi, H. Photovoltaic Panel Parameters Estimation Using an Opposition Based Initialization Particle Swarm Optimization. *Eng. Proc.* **2023**, *29*, 16. [CrossRef]

engineering proceedings

MDPI

Editorial

Statement of Peer Review †

Abdelmadjid Recioui *, Hamid Bentarzi and Fatma Zohra Dekhandji

Laboratory of Signals and Systems, Institute of Electrical and Electronic Engineering, University M'hamed Bougara, Boumerdes 35000, Algeria; hamid.bentarzi@gmail.com (H.B.); fzdekhandji@univ-boumerdes.dz (F.Z.D.)
* Correspondence: a_recioui@univ-boumerdes.dz

† All proceeding papers of this volume are presented at the 2nd International Conference on Computational Engineering and Intelligent Systems, online, 18–20 November 2022.

In submitting conference proceedings to *Engineering Proceedings*, the volume editors of the proceedings certify to the publisher that all papers published in this volume have been subjected to peer review administered by the volume editors. Reviews were conducted by expert referees to the professional and scientific standards expected of a proceedings journal.

- Type of peer review: Single blind
- Conference submission management system: Microsoft CMT
- Number of submissions sent for review: 38
- Number of submissions accepted: 20
- Acceptance rate (number of submissions accepted/number of submissions received): 52.63%
- Average number of reviews per paper: 2
- Total number of reviewers involved: 21
- Peer-review process and regulations: The review process was single-blinded. The main criteria of acceptance were the originality, clarity of the presentation and the findings.

Conflicts of Interest: The authors declare no conflict of interest.

Citation: Recioui, A.; Bentarzi, H.; Dekhandji, F.Z. Statement of Peer Review. *Eng. Proc.* **2023**, *29*, 18. https://doi.org/10.3390/engproc2023029018

Published: 20 June 2023

engineering proceedings

MDPI

Proceeding Paper

Detection and Classification of Power Quality Disturbances Using LSTM †

Fatma Zohra Dekhandji *, Abdelmadjid Recioui, Athmane Ladada and Taha Slimane Moulay Brahim

Laboratory of Signals and systems, Institute of Electrical and Electronic Engineering, University M'hamed Bougara of Boumerdes, Boumerdes 35000, Algeria
* Correspondence: fzdekhandji@univ-boumerdes.dz
† Presented at the 2nd International Conference on Computational Engineering and Intelligent Systems, Online, 18–20 November 2022.

Abstract: The detection and diagnosis of power quality (PQ) issues are critical topics in the electrical power system's generation, transmission, and distribution. Nonlinear loads, power electronic converters, system malfunctions, and switching events are the most common causes of PQ issues. The main purpose of this work is to use an artificial intelligence (AI) technique based on automatic feature extraction to discover and identify PQ difficulties. The AI technique consists of a dedicated architecture of the Long Short-Term Memory (LSTM) network, which is a special type of Recurrent Neural Network (RNN).

Keywords: smart grids; power quality; monitoring; LSTM

1. Introduction

In today's world, power quality (PQ) is important for modern electric power utilities and their consumers, especially when it comes to power quality disturbances, because of the huge increase in nonlinear load, the increased use of sensitive electronic devices, and the need to apply green power globally as well as increase renewable energy applications. Power quality issues have become more crucial than ever [1]. Interruptions and other disturbances are caused by greater usage of semiconductor devices, lighting controls, solid-state switching devices, inverters, and protection and relaying equipment. Such issues have become some of the most pressing concerns for engineers and decision-makers, as frequent occurrences result in significant financial losses for power companies [2].

Detecting power quality issues is critical for improving power quality in electric power systems and for making effective decisions about how to handle network disturbances. Engineers reading waveforms in the field is one method of identifying and classifying power disruptions. Unfortunately, due to the amount of data sampled in current power systems, manual recording is nearly impossible. Furthermore, more equipment failures in power systems occurred because of PQ problems, as well as damage to its sensitive controllers. As a result, the expense of addressing these issues is high in terms of finances, and it results in time losses. Therefore, studying the waveforms of power quality disturbances is important in order to detect and classify them successfully [3]. The power system must overcome various power quality challenges; some of them make it operational, resulting in a blackout of the network. The most significant challenges that the power system must overcome include:voltage spikes, voltage fluctuations, voltage unbalance, voltage sag, voltages swell, harmonic distortion, noise, very short interruptions, and long interruptions [4].

In order to evaluate the power quality and the disturbances that are experienced, it is necessary to detect and classify these disturbances in time, precisely and correctly. To do that, several signal processing methods, such as FT, ST, and WT, have offered sophisticated mathematical algorithms that may detect power quality problems, and these algorithms have shown substantial success in this field [3–5]. Classification techniques like ANN, SVM,

Citation: Dekhandji, F.Z.; Recioui, A.; Ladada, A.; Moulay Brahim, T.S. Detection and Classification of Power Quality Disturbances Using LSTM. *Eng. Proc.* **2023**, *29*, 2. https://doi.org/10.3390/engproc2023029002

Academic Editor: Hamid Bentarzi

Published: 10 January 2023

and Fuzzy logic provide assistance to the detecting algorithm by testing, validating, and training large amounts of signals. This helps to ensure that the selected algorithm is both successful and efficient [6]. Long Short-Term Memory (LSTM) hasprovided a powerful mathematical algorithm for the detection and classification of PQ problems in a power system. LSTM isa developed version of Recurrent Neural Network (RNN). LSTM contains neurons to perform computation (memory cells). These cells have weights and gates; the gates are the defining characteristic of LSTM models. There are three gates within each cell: theinput, forget, and output gates [7].

In this work, the PQ detection and classification system is presented in order to solve part of the power system problems. The paper is divided into the following sections: a description of power quality problems with their causes and effects. Next, a presentation of the existing detection and classification methods is given, with a brief explanation oftheiradvantages and drawbacks. Finally, the simulation results and their discussion are given, with conclusions drawn.

2. Long Short-Term Memory (LSTM)

2.1. Recurrent Neural Networks (RNN)

Recurrent neural networks (RNNs) are networks with an internal network loop that ensures information persistence. RNNs have been particularly developed to process sequential data [8]. RNNs have complete connections between adjacent layers and nodes within the same layer as illustrated in Figure 1. Furthermore, RNNs' hidden units receive feedback from prior states to current status. These properties are appropriate for dealing with temporal–spatial data [9].

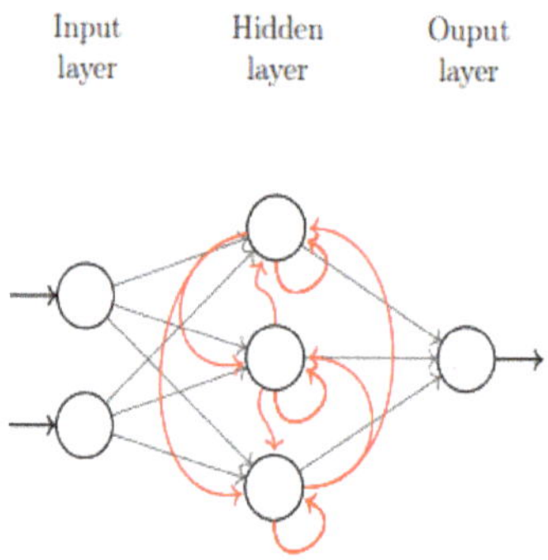

Figure 1. Architecture of a recurrent neural network.

RNNs are referred to as recurrent because they do the same task for each element of a sequence, with the outcome relying on previous estimates. Another way to conceive of RNNs is that they have a "memory" that stores information about previous calculations [10]. As a result, RNNs can link earlier data to the current task. RNN is already being used to solve a number of problems using an internal state (memory) to process a sequence of inputs, including language modeling and translation, speech recognition, text recognition, time series data, and autonomous driving systems to predict vehicle trajectories and assist in avoiding accidents.

2.2. RNN Architecture

In Figure 2, (x_t) denotes the current input of time step t, (h_t) is the output of that time step, and A denotes the network's recurrently connected unit. The previous time step's output is passed on to the following time step as an extra input. If a dense layer is following, the last concealed state (h_t) is passed and processed to form (y_t), or if a dense layer follows, (h_t) is passed and processed to form (y_t).

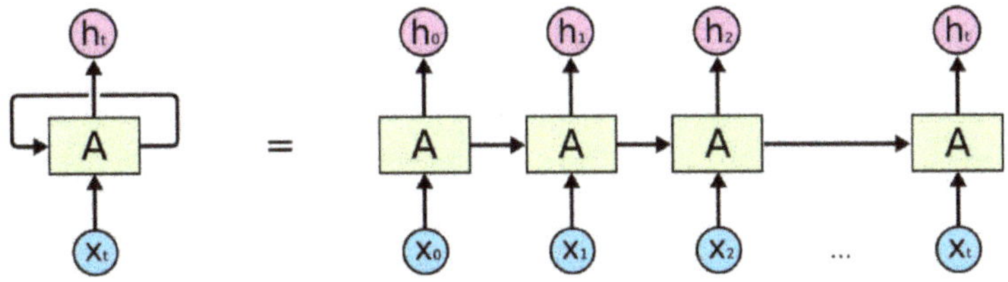

Figure 2. A simple rolled RNN layer and its corresponding unrolled architecture.

2.3. Long Short-Term Memory (LSTM)

Long Short-Term Memory networks are a developed version of the previously mentioned simple RNNs. Hochreiter and Schmidthuber first proposed them in 1997 [11], and they have demonstrated accurate performance in modeling both long- and short-term dependencies of sequential data [8]. In theory, they were created expressly for long-term dependencies in order to overcome the vanishing/exploding gradient problem. Figure 3 shows the structure of a simple LSTM network for prediction, which includes a sequence input layer, an LSTM layer, and a prediction (classification) output layer.

Figure 3. LSTM network system.

2.4. LSTM Network Architecture

The LSTM architecture is based on the existence of a memory cell that is capable of maintaining its state over time and has a nonlinear gating mechanism that controls the flow of information into and out of the cell. The LSTM network is being utilized to create a more complicated and deeper nonlinear neural network capable of demonstrating the effect of long-term memory [12].

LSTM networks have an input layer, an output layer, and a multitude of hidden layers in between. The memory cell is built into the hidden layer. Each cell has three gates (input, forget, and output) and a recurrent connection unit [13].

The LSTM architecture is depicted in Figure 4 as it unfolds over time. The LSTM cell (denoted as A in Figure 4) on the left represents the previous time step, while the cell on the right represents the next time step. The current time step is halfway through. The cell contains three lines. It obtains the input (X_t) and the output from the previous time step in the bottom left corner (the output from the previous layer is called the hidden state in RNNs and is abbreviated as (h_{t-1}). Before running into the four gates denoted as yellow boxes in the drawing, the input (X_t) and the hidden state (h_{t-1}) are concatenated. The cell's third input, which comes from the preceding cell, is represented by a straight arrow that passes through the upper section of the cells. This is the cell state, which allows the LSTM to remember long-term dependencies with a far lower risk of the vanishing and exploding gradient problems that plague typical RNNs [14–17].

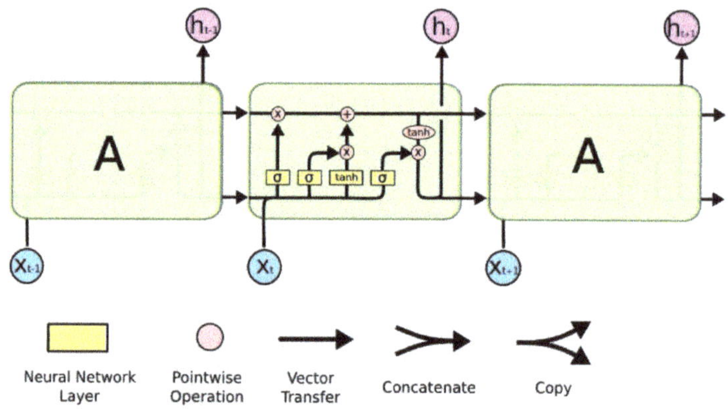

Figure 4. LSTM module containing four interacting layers.

3. Results and Discussions

3.1. Description of the Simulation Environment

MATLAB is used to generate several PQ signals, including interruptions, sag, harmonic distortion, flicker, swell, surge, sag with harmonics, and swell with harmonics, in this section. A comparative study to evaluate the LSTM performance in terms of the detection and classification of these signals is performed. Each simulated waveform is made up of voltage waves recorded at a rate of 64 samples per cycle. A total of 200 case studies are created for each power quality problem by varying the voltage magnitude and the beginning and finish time instants of each PQ problem. Simulated signals are blended with random white noise with signal-to-noise ratios of 40 dB and 20 dB, respectively.

The signal-to-noise ratio (SNR) is a metric that compares the strength of a desired signal to the strength of an additive white Gaussian noise as:

$$\text{SNR} = \frac{\text{Psignal}}{\text{Pnoise}} \tag{1}$$

where P designates the power. Higher values of SNR generally mean a better specification since this means more useful information (the signal) than unwanted data (the noise).

3.2. Proposed Methodology

As shown in Figure 5, the LSTM system is designed and built to perform detection and classification using five parameters: amplitude, start time, end time, duration, and THD percent, as well as two outputs: waveform class and harmonics indicator.

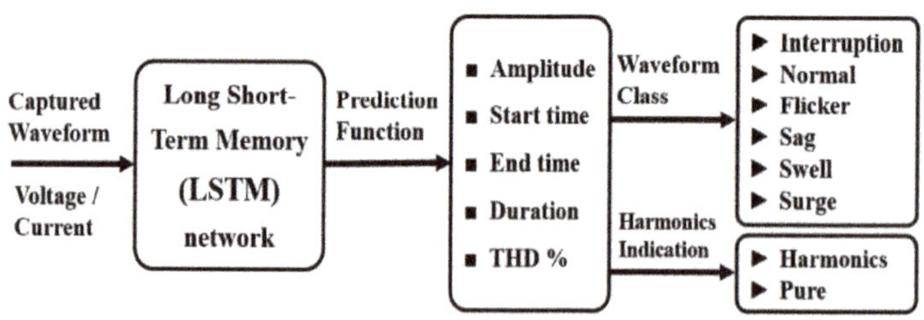

Figure 5. Block diagram of proposed method.

3.3. Model Implementation and Training

In this section, the proposed model is trained by altering the voltage magnitude and the start and finish time instants of each power quality problem; 200 case studies are constructed for each PQ problem. Figure 6 shows the trained system.

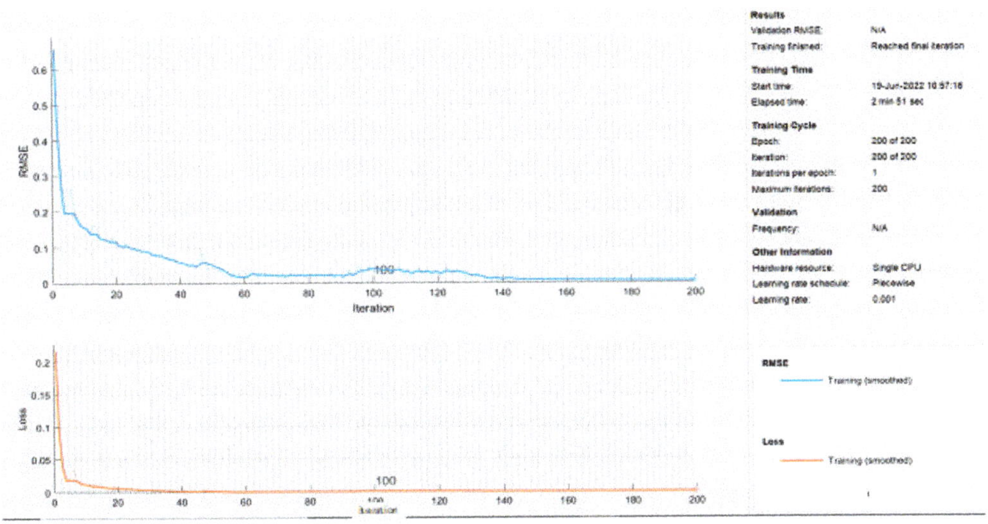

Figure 6. Training process of LSTM model.

To detect and classify the problem waveforms, the LSTM has two outputs as follows:

- The first output of the LSTM function is for the waveform class, which is defined by six sets. These sets are the interruption, sag, normal, flicker, swell, and surge. Any output value, which does not belong to these sets, represents the distortion. The first output of the LSTM network system can assume values between 0 and 3, as shown in Table 1.
- The second output of the LSTM function is for harmonics indication, which is partitioned into two function sets. The labels of these sets are Pure and Harmonics, as shown in Table 1.

Table 1. LSTM Network Outputs.

PQ Problem	LSTM Output
First LSTM output	
Interruption	0
Sag	0.5
Flicker	0.8
Normal sine wave	1
Swell	1.5
Surges	3
Second LSTM output	
Pure	0
Harmonics	1

3.4. Results and Discussions

Table 2 summarizes the performance of the LSTM technique following extensive simulations. The percentage values indicated show how good LSTM is in terms of power quality disturbance detection and classification.

Table 2. Summary of the Obtained results.

Events	Pure Signal	With SNR 40	With SNR 30 dB	With Harmonics	With Harmonics and SNR 30 dB
Normal case	99.4644	99.3551	98.9588	98.9044	95.9687
Sag	99.8875	99.5346	98.2528	99.5662	96.6155
Swell	99.9091	99.3293	97.3609	98.8372	96.7574
Interruption	99.9861	98.5413	97.4985	99.9553	96.1080
Surge	99.4821	97.8860	94.6241	/	/
The totals	99.7458	99.5292	97.3790	99.3158	96.3674

The following remarks may be drawn from Table 2:

- The LSTM has performance values ranging from 96.3674% to 99.7458%.
- A value of 40 dB SNR and the harmonics do not have much of an effect on the detection results, which shows the effectiveness of the LSTM model.
- An SNR of 30 dB is considered a high noise ratio even if the final accuracy was not deduced higher than 3%.
- Merging the normal signal with PQ disturbances at an SNR of 30 dB and harmonics gives an accuracy of around 96%.
- The worst-case PQ disturbances with harmonics and noises were clearly and successfully detected and classified, which proves the LSTM's robustness.

4. Conclusions

The main objective of this work is to develop a Power Quality Detection and Classification using Long Short-Term Memory (LSTM). The method is able to identify and classify several PQ problems (normal, sag, swell, surge, distortion, and interruption) in both simple and complex power quality disturbances in the presence of random noise at different values of the signal-to-noise ratio (SNR). The detection and classification using the LSTM model aredesigned and implemented to perform with five input data parameters (amplitude, start time, end time, duration, THD %) and two outcomes: the waveform class and a harmonics content indicator.The detection and classification of PQ problems using LSTM show the high accuracy of this model, even in the worst cases of adding harmonics with 30-dB SNR simultaneously with PQ problemsignals. This proves the robustness of the LSTM method, as the obtained classification rate was 96.3674%. Hence, the findings of the present work prove that the LSTM network is a leading method in the world of PQ detection and classification.

Author Contributions: Conceptualization, F.Z.D. and A.R.; methodology, A.R.; software, F.Z.D.; validation, A.R., T.S.M.B. and A.L.; formal analysis, F.Z.D.; investigation, A.R.; resources, F.Z.D.; data curation, T.S.M.B.; writing—original draft preparation, F.Z.D.; writing—review and editing, A.R.; visualization, F.Z.D.; supervision, A.R.; project administration, F.Z.D. All authors have read and agreed to the published version of the manuscript.

Funding: This research received no external funding.

Institutional Review Board Statement: Not applicable.

Informed Consent Statement: Not applicable.

Data Availability Statement: The data presented in this study are available on request from the corresponding author. The data have been synthesized for the purpose of this study and are not publicly available.

Conflicts of Interest: The authors declare no conflict of interest.

References

1. Lei, J.; Liu, C.; Jiang, D. Fault diagnosis of wind turbine based on Long Short-term memory networks. *Renew. Energy* **2019**, *133*, 422–432. [CrossRef]
2. Dekhandji, F.Z.; Rais, M.C. A Comparative Study of Power Quality Monitoring Using Various Techniques. In *Optimizing and Measuring Smart Grid Operation and Control;* IGIGlobal: Hershey, PA, USA, 2021; pp. 259–288.
3. Dekhandji, F. Signal Processing Deployment in Power Quality Disturbance Detection and Classification. *Acta Phys. Pol. A* **2017**, *132*, 415–419. [CrossRef]
4. Dekhandji, F.Z.; Talhaoui, S.; Arkab, Y. Power Quality Detection, Classification and Monitoring Using LABVIEW. *Alger. J. Signals Syst.* **2019**, *4*, 101–111. [CrossRef]
5. Dekhandji, F.Z.; Bentarzi, H.; Recioui, A. Effects of voltage disturbances on induction motors and adjustable speed drives. In Proceedings of the 2015 4th International Conference on Electrical Engineering (ICEE), Boumerdes, Algeria, 29–31 October 2015; pp. 1–4.
6. Dekhandji, F.Z.; Douche, M.; Zebidi, N. DVR and D-STATCOM Mitigation Techniques of Power Quality Effects on Induction Motors. *Alger. J. Signals Syst.* **2017**, *2*, 110–129. [CrossRef]
7. Balouji, E.; Gu, I.Y.; Bollen, M.H.; Bagheri, A.; Nazari, M. ALSTM-based deep learning method with application to voltage dip classification. In Proceedings of the 2018 18th International Conference on Harmonics and Quality of Power (ICHQP), Ljubljana, Slovenia, 13–16 May 2018; pp. 1–5.
8. Rens, J.; DeKock, J.; Van Wyk, W.; Van Zyl, J. The effect of real network phase disturbances on the calculation of IEC61000-4-30 parameters. In Proceedings of the 2014 16th International Conference on Harmonics and Quality of Power (ICHQP), Bucharest, Romania, 25–28 May 2014; pp. 303–306.
9. De Almeida, P.R.; da Silva, A.M.; Lemes, M.I.; Garcia, F.R. Application of reactive compensation equipment in industrial systems under aspects of harmonics and switching transients: A study of real case. In Proceedings of the 2016 17th International Conference on Harmonics and Quality of Power (ICHQP), Belo Horizonte, Brazil, 16–19 October 2016; pp. 798–803.
10. Langella, R.; Testa, A.; Et, A. *IEEE Recommended Practice and Requirements for Harmonic Control in Electric Power Systems;* IEEE: Piscataway, NJ, USA, 2014.
11. Jaekel, B.W. Description and classification of electromagnetic environments-revision of IEC61000-2-5. In Proceedings of the 2008 IEEE International Symposium on Electromagnetic Compatibility, Detroit, MI, USA, 18–22 August 2008; pp. 1–4.
12. De, S.; Debnath, S. Real-time cross-correlation-based technique for detection and classification of power quality disturbances. *IETGener. Transm. Distrib.* **2017**, *12*, 688–695. [CrossRef]
13. Babu, N.R.; Mohan, B.J. Fault classification in power systems using EMD and SVM. *Ain Shams Eng. J.* **2017**, *8*, 103–111. [CrossRef]
14. Zhang, S.; Wang, Y.; Liu, M.; Bao, Z. Data-Based Line Trip Fault Prediction in Power Systems Using LSTM Networks and SVM. *IEEE Access* **2018**, *6*, 7675–7686. [CrossRef]
15. Dekhandji, F.Z. Detection of power quality disturbances using discrete wavelet transform. In Proceedings of the 2017 5th International Conference on Electrical Engineering-Boumerdes (ICEE-B), Boumerdes, Algeria, 29–31 October 2017; pp. 1–5.
16. Method Based on Long Short-Term Memory Neural Networks. In Proceedings of the 2017 IEEE 86th Vehicular Technology Conference (VTC-Fall), Toronto, ON, Canada, 24–27 September 2017; pp. 1–5.
17. Tsiouris, K.M.; Pezoulas, V.C.; Zervakis, M.; Konitsiotis, S.; Koutsouris, D.D.; Fotiadis, D.I. A LongShort-Term Memory deep learning network for the prediction of epileptic seizures using EEG signals. *Comput. Biol. Med.* **2018**, *99*, 24–37. [CrossRef] [PubMed]

engineering proceedings

MDPI

Proceeding Paper

Opportunistic Load Balancing for Virtual Machines Scheduling in a Cloud Environment †

Younes Khair [1,*] and Haouari Benlabbes [2]

1 Department of Computer Science, Faculty of Exact Sciences, University of Tahri Mohammed, Bechar 08000, Algeria
2 Department of Exact Sciences, Higher Normal School of Bechar, Bechar 08000, Algeria
* Correspondence: ynss.khair@univ-bechar.dz
† Presented at the 2nd International Conference on Computational Engineering and Intelligent Systems, Online, 18–20 November 2022.

Abstract: One of the Internet's fastest-growing technologies is cloud computing, which needs each user to share various resources and virtual computers to access the services they demand. The difficulty is in allocating the user's tasks among the many resources at hand while maintaining a balanced workload. In this work, we present the design of a new scheduling approach for cloud virtual machines in order to process events (related to resource overuse or underuse). This reduces the amount of effort consumed while maintaining the required performance levels in the cloud data center. To evaluate the effectiveness of the approach with real load paths, we conducted an experiment using the open-source cloud computing platform OpenNebula.

Keywords: cloud computing; virtual machine scheduling; opportunistic load balancing

Citation: Khair, Y.; Benlabbes, H. Opportunistic Load Balancing for Virtual Machines Scheduling in a Cloud Environment. *Eng. Proc.* **2023**, *29*, 1. https://doi.org/10.3390/engproc2023029001

Academic Editors: Abdelmadjid Recioui, Hamid Bentarzi and Fatma Zohra Dekhandji

Published: 10 January 2023

1. Introduction

Cloud computing has emerged in recent years as a new area of information technology. According to the National Institute of Standards and Technology, cloud computing is a model that allows users to access on demand and from anywhere on the Internet, to a shared pool of computing resources that can be provisioned quickly and flexibly, with minimal administrative effort and interaction with the service provider [1].

Virtualization technology is used in data centers to enable the existence of several virtual machines on the same physical machine. This capacity to personalize allows the service provider to maximize the investment of available physical resources, resulting in greater profits [2].

Cloud resource management activities involve examining the resources accessible in the cloud environment, offering access to end users, monitoring security, managing resource reservations, and monitoring all cloud layers. Because resource allocation management has a significant impact on the performance and efficiency of the cloud, as well as the cost of its users, it has become a difficult challenge to manage the various cloud resources, particularly to ensure cloud computing performance through strategic planning techniques that maintain cloud performance. This performance may be accomplished by maintaining a balanced distribution of workloads across cloud resources. One of the most challenging difficulties in cloud computing is finding a suitable level of task distribution among the many resources used in the cloud, namely: (one or more servers, hard drives, network links, or other resources). Thus, cloud service providers provide a means for distributing application requests to any number of data center-based application propagations. A balanced distribution aims to enhance processing speed while decreasing reaction time [3–5].

2. Related Work

The significant expansion of cloud computing's technology offerings is a reflection of both the system's user base and the tasks those users are requiring it to perform. As a result, a significant number of scheduling algorithms were developed and used in different cloud computing environments. The performance of these algorithms was then assessed based on a variety of different criteria [6].

Researchers have recently shown a great interest in scheduling in cloud computing. Several scheduling methods operating inside the cloud computing system have been suggested by various researchers.

For the purpose of scheduling distinct tasks, several heuristic algorithms have been developed and put to use in the cloud environment. The First Come First Serve basis algorithm, Min-Max algorithm, Min-Min algorithm, and Suffrage algorithm are among the most significant developments in heuristic algorithms. Other important developments include Greedy Scheduling, Shortest Task First, Sequence Scheduling, Balance Scheduling (BS), Opportunistic Load Balancing, and Min-Min Opportunistic Load Balancing [2–5,7].

In this work, we focus on using an algorithm that, for scheduling based on load balancing, offers the best performance in the cloud computing system. For a set of virtual machines, the algorithm was evaluated in two different scenarios. We looked into the subsequent performance scenarios: (1) the effect of load balancing throughput gain; (2) the effect of resource utilization rate.

3. OpenNebula

It is an open-source platform with an Apache 2 license that aims to develop a common cloud computing solution for controlling large and distributed infrastructures. Additionally, it has more capabilities, more adaptable methods, and enhanced cloud formation interoperability. OpenNebula offers hybrid clouds, which offer extremely flexible hosting options while integrating on-premises and public cloud architecture [6].

OpenNebula is currently on version (6.3.9). Its design is flexible, allowing interaction with different types of storage, network infrastructure, and hypervisor technologies. Figure 1 shows a diagram with the components grouped into three layers: (i) Drivers, (ii) Core, and (iii) Tools. These elements connect to each other through a set of APIs that provide system and cloud user interfaces [6].

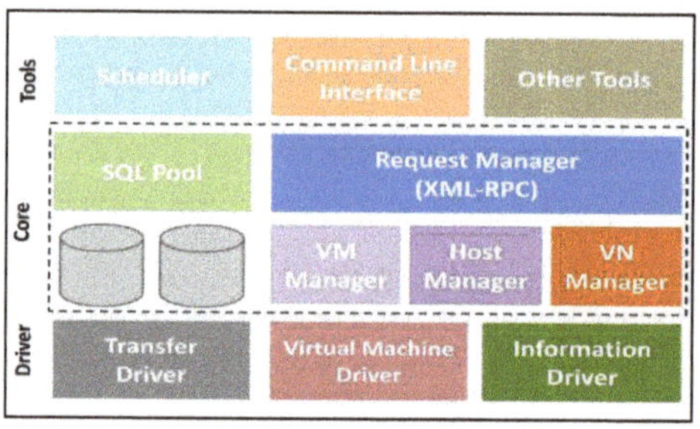

Figure 1. OpenNebula Architecture [8].

4. Opportunistic Load Balancing for Scheduling Virtual Machines in the Cloud Environment

The objective of this algorithm is to assign the work to the cloud host machine that is under the least increase in load when compared to other cloud hosts. The aim is to first identify each host machine's load distribution before allocating virtual machines to it. Then, the machine with the lower load level is chosen to perform the required function or task [9].

In our architecture concept (Figure 2), the opportunistic load balancing OLB is in charge of scheduling when these activities will be performed out on the pre-activated virtual machines. Hosts have two distinct resources (CPU and RAM). Each request for allocation consists of two values that describe the virtual machine's execution requirements for CPU and RAM on one of the Cloud hosts. Additionally, OLB communicates with OpenNebula, using the OpenNebula Cloud API (OCA) which encapsulates access to OpenNebula's XML-RPC API and implicitly associated functions as shown in the definition in Section 3 [6]. To establish the connection, the OpenNebula XML-RPC address access methods are chosen and assigned. This integration allows OpenNebula to run OBL queries.

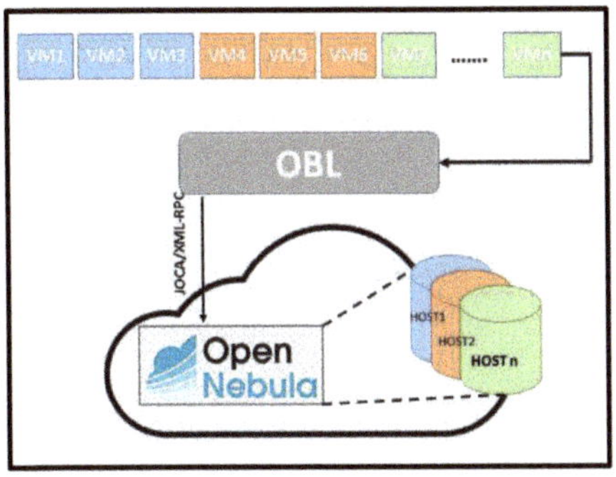

Figure 2. The architecture concept.

4.1. Scheduling Based on Opportunistic Load Balancing

Suppose the fact that there are $|N|$ similar resource host machines at the beginning, where H = {1, 2, . . . , H} The capacity of each host machine is Ci where $i \in$ H. Suppose that the capacities are given integer values, where each integer roughly corresponds to the CPU and RAM requirements of the host machine. Consider three machines, each hosting a distinct load from the others, H1 of C1 = 10%, H2 of C2 = 50%, and H3 of C3 = 90%, as illustrated in Figure 3.

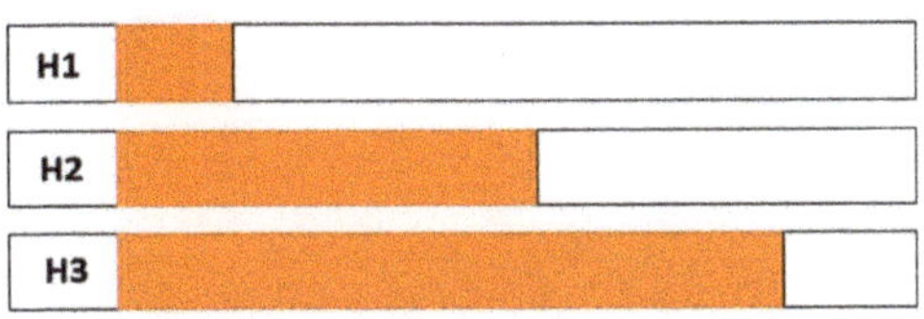

Figure 3. An initial situation for load balancing.

The capacity is proved by:

$$Ci = \frac{CPUuse * RAMuse}{100}$$

Consider that there are $|N|$ virtual machines, $V = \{1, 2, \ldots, V\}$, each of which has a weight of $W = \{W1, W2, \ldots\ Wn\}$ and an end time for the activity that is being executed on it of T = {T1, T2, . . . Tn}, The virtual machine's weight is a representation of the funding necessary for its implementation. Whenever it uses a percentage of the host's resources, the weight is defined as follows:

$$Wi = \frac{CPUreq * RAMreq}{100}$$

To maximize the value of all virtual machines and ensure that the overall weight of virtual machines in each resource does not exceed that resource's capacity, the challenge of allocating virtual machines to resources occurs. Let u_{ij} The value reflects how significant this virtual machine is for the resource.

$$\sum_{i=1}^{H}\sum_{j=1}^{V} u_{ij}T_{ij}$$

4.2. The Proposed Scenario

Through the scenario (without and with the use of opportunistic load balancing), this algorithm was examined regarding the cost factors and rate of resource utilization. The parameter values for the scenario's results are summarized in Table 1.

Table 1. Specifications of the environment used for the experiments.

Host machines	Number of hosts	3
	RAM	2 GB
	Storage	250 GB
Virtual Machines	Number of Virtual Machines	10
	RAM	256 MB
	Number of processors	1

During the evaluation of the proposed planning algorithm, the characteristics of all the created virtual machines were installed and harmonized in the homogenous environment.

4.2.1. Load Balancing Throughput Gain

This is the average success rate for allocations on host machines. For each virtual machine that is allocated, the allocation rate of incoming tasks is calculated, and data is supplied at the end of each phase of allocation. The higher this factor, the better the algorithm performs. The value of the throughput gain is given as follows:

$$G_t = \sum_{alloc^i} (Exe_{time})$$

4.2.2. Resource Use Rate

The ratio of the host machine's total occupation to the totality of all the virtual machines' execution times is referred to as the resource utilization rate (in our scenario, it is the time taken to complete the execution).

$$Usage\ rate\ of\ each\ host\ machine = \frac{\begin{matrix} The\ final\ time\ for\ the\ completion \\ of\ work\ on\ the\ host\ machine \end{matrix}}{Total\ runtime\ for\ all\ virtual\ machines} * 100$$

5. Results and Discussion

5.1. Load Balancing Throughput Gain

The behavior of throughput as a function of virtual machines in the what-if scenarios is shown in Figure 4. The results show that when a single resource is used, performance degrades as the number of allowances increase as load balancing is not enabled (there is no OBL). In this case, the host machines are not taken into account in terms of virtual machine allocation. In the opportunistic load-balancing scenario, average productivity returns to maximum levels despite an increase in the number of virtual machines.

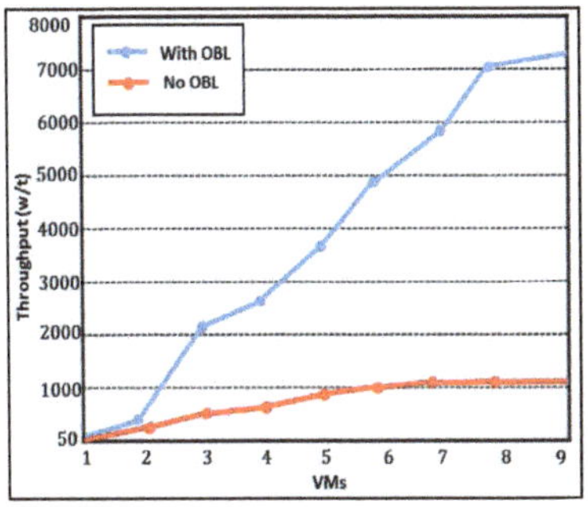

Figure 4. The behavior of throughput as a function of virtual machines.

5.2. Resource Use Rate

As we previously mentioned, the rate of resource utilization refers to the correlation between the overall host machine occupation and the execution timings of all virtual machines. Figure 5 shows that opportunistic load balancing has the best resource utilization rate.

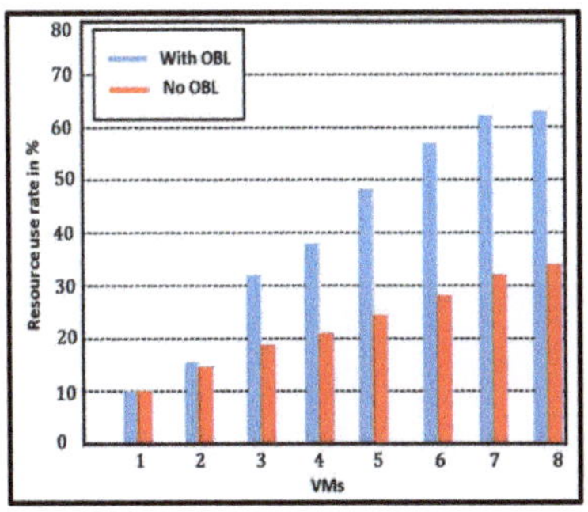

Figure 5. Resource use rate.

6. Conclusions

Through this research, we have planned to establish an algorithm that, for scheduling based on load balancing, offers the best performance in the cloud computing system. For a set of virtual machines, the algorithm was evaluated in two different scenarios. We looked into the subsequent performance scenarios: (1) the effect of load balancing throughput gain; (2) the effect of resource utilization rate.

Author Contributions: Conceptualization, H.B. and Y.K.; methodology, Y.K.; software, Y.K.; validation, Y.K and H.B.; formal analysis, Y.K.; investigation, Y.K.; resources, Y.K.; data curation, Y.K.; writing—original draft preparation, Y.K.; writing—review and editing, Y.K.; visualization, Y.K.; supervision, H.B.; project administration, H.B.; funding acquisition, Y.K. All authors have read and agreed to the published version of the manuscript.

Funding: This research received no external funding.

Institutional Review Board Statement: Not applicable.

Informed Consent Statement: Not applicable.

Data Availability Statement: No new data were created or analyzed in this study. Data sharing is not applicable to this article.

Conflicts of Interest: The authors declare no conflict of interest.

References

1. William, V.; Broberg, J.; Buyya, R. Introduction to cloud computing. *Cloud Comput. Princ. Paradig.* **2011**, 1–44. [CrossRef]
2. Qi, Z.; Cheng, L.; Boutaba, R. Cloud computing: State-of-the-art and research challenges. *J. Internet Serv. Appl.* **2010**, *1*, 7–18.
3. Pietri, I.; Chronis, Y.; Ioannidis, Y. Multi-objective optimization of scheduling dataflows on heterogeneous cloud resources. In Proceedings of the 2017 IEEE International Conference on Big Data (Big Data), Boston, MA, USA, 11–14 December 2017; pp. 361–368.
4. Sofia, A.S.; Kumar, P.G. Multi-objective task scheduling to minimize energy consumption and makespan of cloud computing using NSGA-II. *J. Netw. Syst. Manag.* **2018**, *26*, 463–485. [CrossRef]
5. Elaziz, M.A.; Xiong, S.; Jayasena, K.; Li, L. Task scheduling in cloud computing based on hybrid moth search algorithm and differential evolution. *Knowl.-Based Syst.* **2019**, *169*, 39–52. [CrossRef]
6. Khair, Y.; Dennai, A.; Elmir, Y. Dynamic and elastic monitoring of VMs in cloud environment. *J. Supercomput.* **2022**, *78*, 19114–19137. [CrossRef]
7. Deafallah, A. A metaheuristic framework for dynamic virtual machine allocation with optimized task scheduling in cloud data centers. *IEEE Access* **2021**, *9*, 74218–74233.
8. Younes, K.; Dennai, A.; Elmir, Y. An Experimental Performance Evaluation of OpenNebula and Eucalyptus Cloud Platform Solutions. In *International Conference on Artificial Intelligence in Renewable Energetic Systems;* Springer: Cham, Germany, 2021.
9. Minxian, X.; Tian, W.; Buyya, R. A survey on load balancing algorithms for virtual machines placement in cloud computing. *Concurr. Comput. Pract. Exp.* **2017**, *29*, e4123.

engineering proceedings

MDPI

Proceeding Paper

Observer Backstepping Design for Flight Control †

Ben Messaoud Safinaz *, Belkheiri Mohammed and Belkheiri Ahmed

Laboratory of Telecommunications, Signals and Systems, University Amar Telidji of Laghouat, P.O. Box 37G, Guardaia Road, Laghouat 03000, Algeria
* Correspondence: s.benmessaoud@lagh-univ.dz
† Presented at the 2nd International Conference on Computational Engineering and Intelligent Systems, Online, 18–20 November 2022.

Abstract: This paper presents observer backstepping as a new nonlinear flight control design framework. Flight control laws for general-purpose maneuvering in the presence of nonlinear lift and side forces are designed. The controlled variables are the angle of attack, the sideslip angle, and the roll rate. The stability has been proved using Lyapunov stability criteria. Control laws were evaluated using realistic aircraft simulation models, with highly encouraging results.

Keywords: observer backstepping; stability; control; Lyapunov function; fighter aircraft; maneuvering

Citation: Ben Messaoud, S.; Belkheiri, M.; Belkheiri, A. Observer Backstepping Design for Flight Control. *Eng. Proc.* **2023**, *29*, 3. https://doi.org/10.3390/engproc2023029003

Published: 10 January 2023

1. Introduction

Several novel design approaches for controlling nonlinear dynamic systems have been developed in recent years. Backstepping is one of these techniques. Backstepping allows for the methodical design of stabilizing control laws in certain classes of nonlinear systems. A backstepping design can be used in systems with limited model information. In this scenario, the aim is to choose a control law that decreases the Lyapunov function for all systems with the given model uncertainty. As a result of its simultaneous and often implicit appearance in several papers in the late 1980s, the origin of backstepping is not quite clear. The work of Professor Petar V. Kokotovi´c and his coworkers has, however, brought backstepping to the forefront to a significant extent. The 1991 Bode lecture at the IEEE CDC, held by Kokotovi´c [1], was devoted to the evolving subject and in 1992, Kanellakopoulos et al. [2] presented a mathematical "toolkit" for designing control laws for various nonlinear systems using backstepping, where backstepping has to be performed under state observation. Backstepping designs are available for a wide range of electrical motors [3–6]. Wind turbines are considered in [7], ref [8] using backstepping while observer backstepping is the subject of [9]. In [10,11], backstepping is used for automatic ship positioning. Robotics is another field where backstepping designs can be found. Tracking control is considered in [12] and [13]. Backstepping control augmented by neural networks is proposed to address the tracking problem for robot manipulators [14] and for an induction machine based on a modified (FOC) method [15]. There are also a few papers that combine flight control and backstepping [16,17]. Ref. [18] treats formation flight control of unmanned aerial vehicles. Refs. [19,20] use backstepping to design flight control laws which are adaptive to changes in the aerodynamic forces and moments due to, e.g., actuator failures.

The reviewed literature reveals a key common feature, which is complete knowledge of the system states. In cases where not all the state variables can be measured, the need for observers arises. In general, the separation principle valid for linear systems does not hold for nonlinear systems in general. Therefore, care must be taken when designing the feedback law based on the state estimates. This is the topic of observer backstepping [2,21]. Comparatively, with the advances in nonlinear control theory, there is a desire in aircraft technology to achieve supermaneuverability for fighter aircraft. Using high angles of attack can lead to tactical advantages, as Herbst [22] and Well et al. [23] demonstrate when evaluating aircraft reversal maneuvers. The target is for the aircraft to return to its original position of departure with the same speed and altitude but on the opposite heading in the shortest time possible. By applying high angles of attack during the turn, the aircraft can maneuver in less airspace and complete the maneuver in less time. Flight dynamics are not linear in these maneuvers. Due to this change, linear control design tools are no longer useful for flight control design. In this paper, we investigate how backstepping can be used for flight control design to achieve stability over the entire flight envelope. Control laws based on the state estimates for general-purpose maneuvering are derived and their properties are investigated.

The paper is organized in the following manner: Section 2 presents the dynamic model of the aircraft. Section 3 describes the controller design methodology, which presents a backstepping observer design. Section 4 illustrates the simulation results and discussion of the proposed controller. The control strategy was applied using a realistic aircraft simulation model. Finally, conclusions are given in Section 5.

2. Aircraft Dynamics

Figure 1 illustrates the controlled variables, which are the angle of attack, sideslip angle, and roll rate about the stability x-axis.

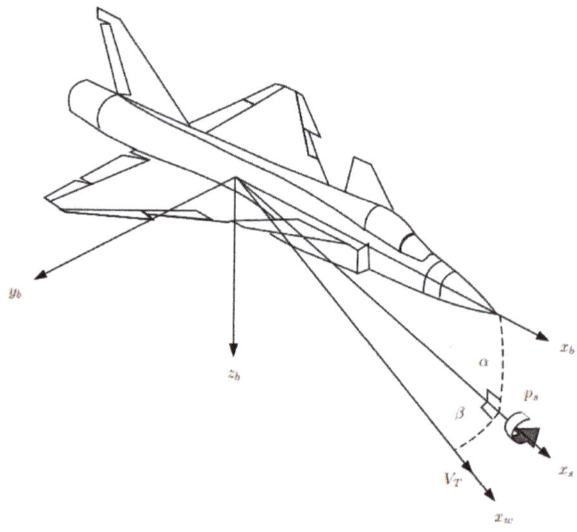

Figure 1. Lateral control objectives.

The equations linking the variables to be controlled to control inputs are presented in Stevens and Lewis [24] and Boiffier's [25] books. Who focuses on developing a model for control design, which consists of first-order differential equations.

$$\dot{\alpha} = q - (p\cos\alpha + r\sin\alpha)\tan\beta + \frac{1}{mV_T\cos\beta}(-L - F_T\sin\alpha + mg_1) \quad (1)$$

$$\dot{\beta} = p\sin\alpha - r\cos\alpha + \frac{1}{mV_T}(Y - F_T\cos\alpha\sin\beta + mg_2) \quad (2)$$

m is the aircraft mass, F_T is the engine thrust force and V_T is the total velocity. L and Y are the lift and side forces respectively.

Figure 2 shows the typical lift force and side force coefficients used for the simulations later.

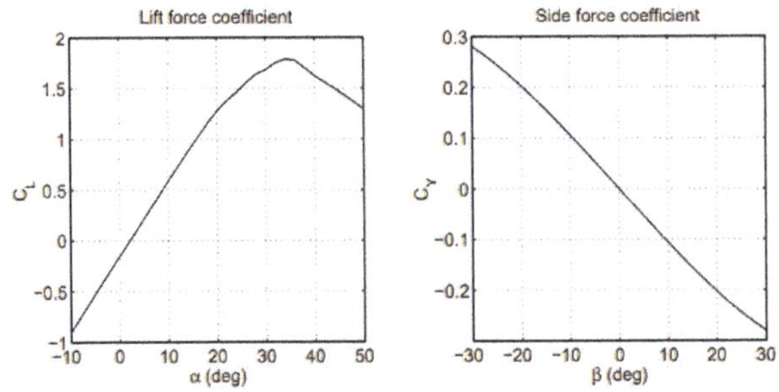

Figure 2. Typical lift force coefficient vs. angle of attack and side force coefficient vs. sideslip characteristics.

The force contributions due to gravity depend on the orientation of the aircraft, given by the pitch angle, θ, and the roll angle, ϕ, are

$$g_1 = g(\cos\alpha\cos\theta\cos\phi + \sin\alpha\sin\theta)$$
$$g_2 = g(\cos\beta\cos\theta\sin\phi + \sin\beta\cos\alpha\sin\theta - \sin\alpha\sin\beta\cos\theta\cos\phi)$$

Since the roll rate to be controlled, p_s, is expressed in the stability-axes coordinate system, we need to establish the relationship between the body-axes angular velocity

$$\omega = \begin{pmatrix} p & q & r \end{pmatrix}^T$$

To the stability-axes angular velocity,

$$\omega_s = \begin{pmatrix} p_s & q_s & r_s \end{pmatrix}^T$$

through the transformation

$$\omega_s = R_{sb}\omega \tag{3}$$

where

$$R_{sb} = \begin{pmatrix} \cos\alpha & 0 & \sin\alpha \\ 0 & 1 & 0 \\ -\sin\alpha & 0 & \cos\alpha \end{pmatrix}$$

In the control design, it is more convenient to work with ω_s rather than ω. Introducing

$$u = \begin{pmatrix} u_1 & u_2 & u_3 \end{pmatrix}^T = \dot{\omega}_s \tag{4}$$

Then we can rewrite the aircraft dynamics (1)–(2) as

$$\dot{p}_s = u_1 \tag{5}$$

$$\dot{\alpha} = q_s + f_\alpha \tag{6}$$

$$\dot{q}_s = u_2 \tag{7}$$

$$\dot{\beta} = -r_s + f_\beta \tag{8}$$

$$\dot{r}_s = u_3 \tag{9}$$

With

$$f_\alpha = -p_s tan\beta + \frac{1}{mV_T cos\beta}(-L(\alpha) - F_T sin\alpha + mg_2)$$
$$f_\beta = \frac{1}{mV_T}(Y(\beta) - F_T cos\alpha sin\beta + mg_3)$$

The lift and side force coefficient, C_L and C_Y, mainly depend on α and β respectively [26], this yields

$$L(\alpha) = \bar{q}SC_L(\alpha)$$
$$Y(\beta) = \bar{q}SC_Y(\beta)$$

where $\bar{q} = \frac{1}{2}\rho(h)V_T^2$ is the aerodynamic pressure, ρ is the air density, and S is the wing plan form area.

3. Observer Backstepping Design for Flight Control

3.1. Control Objective

Controlling maneuverability is the main objective in a dogfight. In order to control maneuverability, you must include relevant controlled variables in the lateral and longitudinal directions. Assuming that the latter directions are not applied simultaneously. We only deal with lateral commands in this paper, therefore the control objective is for the angle of attack α and the stability axis roll rate p_s to follow α^{ref} and p_s^{ref}, respectively. With the standard assumption being that a roll is performed with a steady angle of attack and no sideslip β. The sideslip must be kept at zero at all times in order for the aircraft to fly in a straight line. Furthermore, the aircraft velocity, V_T and Euler angles, ϕ, θ, and ψ are considered constant.

The observer backstepping approach used in flight control laws based on the state estimates for general-purpose maneuvering does not cancel out the nonlinear parts of lift and side forces, $L(\alpha)$ and $Y(\beta)$, respectively.

3.2. Control Design

As a starting point, note that the dynamics of the angle of attack (6)–(7) are structurally similar to the sideslip dynamics (8)–(9). The two second-order systems can be expressed as

$$\begin{aligned} \dot{\eta}_1 &= f(\eta_1) + \eta_2 \\ \dot{\eta}_2 &= u \end{aligned} \tag{10}$$

where $\eta = \begin{bmatrix} \eta_1 & \eta_2 \end{bmatrix}^T = \begin{bmatrix} \alpha & q_s \end{bmatrix}^T$ Then we can rewrite (10) as

$$\begin{aligned} \dot{\eta} &= A\eta + \sigma f(y) + Bu \\ y &= C\eta \end{aligned} \tag{11}$$

where

$$A = \begin{bmatrix} 0 & 1 \\ 0 & 0 \end{bmatrix}, \quad B = \begin{bmatrix} 0 \\ 1 \end{bmatrix},$$
$$C = \begin{bmatrix} 1 & 0 \end{bmatrix}, \quad \sigma = \begin{bmatrix} 1 \\ 0 \end{bmatrix}$$

A measurement can be made only with y. The following observer can be used to estimate the state of this class of nonlinear systems, characterized by geometric conditions in [[27], Theorem 5.1].

$$\dot{\hat{\eta}} = A\hat{\eta} + K_0(\hat{\eta}_1 - y) + \sigma f(y) + Bu \tag{12}$$

The error system is

$$\dot{\tilde{\eta}} = A_0\tilde{\eta}, \quad \tilde{\eta} = \eta - \hat{\eta} \tag{13}$$

K_0 is chosen so that $A_0 = A - K_0C$ in (13) is Hurwitz.

We only specify one subsystem, $\alpha - dynamics$ (1), and the other subsystem, $\beta - dynamics$ (2) should be identical to the latter specified, for the derivation of observer backstepping.

Step 1: We define the output error $z_1 = \eta_1 - \eta_1^{ref} = \eta_1 - \lambda_0(\eta_1^{ref})$ and consider the second-order system

$$\dot{z}_1 = \eta_2 + f(\eta_1) - \dot{\eta}_1^{ref} \triangleq \eta_2 + \mu_1(\eta_1, \eta_1^{ref}, \dot{\eta}_1^{ref}) \quad (14)$$

$$\dot{\eta}_2 = u_2 \quad (15)$$

Similarly to Lemma OIB [2], the following system is composed by replacing (15) with the second equation of the observer (12). With $\tilde{\eta}_2 = \eta_2 - \hat{\eta}_2$

$$\dot{z}_1 = \hat{\eta}_2 + \mu_1(\eta_1, \eta_1^{ref}, \dot{\eta}_1^{ref}) + \tilde{\eta}_2 \quad (16)$$

$$\dot{\hat{\eta}}_2 = u_2 + K_{02}(\hat{\eta}_1 - \eta) \quad (17)$$

In the spirit of backstepping, we start by regarding $\hat{\eta}_2$ as the control input of (16)–(17), then Lemma NDM [2] would result in

$$\hat{\eta}_2^{des} = -\mu_1(\eta_1, \eta_1^{ref}, \dot{\eta}_1^{ref}) - c_\alpha z_1 \triangleq \lambda_1(\eta_1, \eta_1^{ref}, \dot{\eta}_1^{ref}) \quad (18)$$

and

$$V_1(z_1, \tilde{\eta}) = \frac{1}{2}z_1^2 + \tilde{\eta}^T P_0 \tilde{\eta} \quad (19)$$

where

$$P_0 A_0 + A_0^T P_0 = -I$$

Step 2: Defining the state error $z_2 = \hat{\eta}_2 - \lambda_1(\eta_1, \eta_1^{ref}, \dot{\eta}_1^{ref})$ and (16)–(17) is rewritten as

$$\dot{z}_1 = z_2 - c_\alpha z_1 + \tilde{\eta}_2 \quad (20)$$

$$\dot{z}_2 = u_2 + K_{02}(\hat{\eta}_1 - \eta) - \frac{\partial \lambda_1}{\partial \eta_1}\dot{\eta}_1 - \frac{\partial \lambda_1}{\partial \eta_1^{ref}}\dot{\eta}_1^{ref} - \frac{\partial \lambda_1}{\partial \dot{\eta}_1^{ref}}\ddot{\eta}_1^{ref} \triangleq u_2 + \mu_2(\eta_1, \hat{\eta}_1, \hat{\eta}_2, \eta_1^{ref}, \dot{\eta}_1^{ref}, \ddot{\eta}_1^{ref}) - \frac{\partial \lambda_1}{\partial \eta_1}\tilde{\eta}_2 \quad (21)$$

We select the control law based on Lemmas OIB and NDM [2]

$$u_2 = -\mu_2(\eta_1, \hat{\eta}_1, \hat{\eta}_2, \eta_1^{ref}, \dot{\eta}_1^{ref}, \ddot{\eta}_1^{ref}) - [1 + (\frac{\partial \lambda_1}{\partial \eta_1})^2]z_2 - z_1 \quad (22)$$

and

$$V_2(z_1, z_2, \tilde{\eta}) = V_1(z_1, \tilde{\eta}) + \frac{1}{2}z_2^2 + \tilde{\eta}^T P_0 \tilde{\eta} \quad (23)$$

The derivative of (23) along the solutions of (20)–(22) is nonpositive

$$\dot{V}_2 \leq 0$$

With this methodical approach, we designed the control law (22), and we created the conditions for the results mentioned in [2].

Controlling the stability axis roll p_s is straightforward. Given the dynamics from (5) and the roll rate command p_s^{ref}, simply assign

$$u_1 = c_{p_s}(p_s^{ref} - p_s) \quad (24)$$

where $c_{p_s} > 0$.

4. Simulation Results and Discussion

In the simulations, an Admire, a MATLAB/SIMULINK environment for the (GAM) [28] is considered with properties summarized in Table 1.

Table 1. GAM Properties.

Entity	Values
Mass (m)	9100 (kg)
Wing planform area (S)	45 (m)
Density of the airflow (ρ)	1.08 (kg/m^3)

A flight scenario in which the aircraft is in level flight at Mach 0.5 and a height of 1000 m was chosen for the simulations. The simulations included actuator and sensor dynamics that were not included in the previous design. Additionally, Euler angles were given such small values. The total velocity ranges between 100 and 170 km/h, with $F_T = 500$ N.

For the control law parameters were set $c_\alpha = 75, c_\beta = 50, c_{p_s} = 25$ for the backstepping approach and $K_{02} = \frac{\gamma}{\epsilon^2}$ for the second equation of the observer, where $\gamma = 1$ and $\epsilon = 0.001$. Figure 3 illustrates the trajectory tracking simulation results for such an angle of attack, slideslip, and stability axis roll rate, and it is obvious that each of the latter variable graphs properly follows the prescribed path. Both α and β are shown in Figure 4 with their estimates, resulting in a perfect estimation of the observer chosen. For the control input, it is presented in Figure 5.

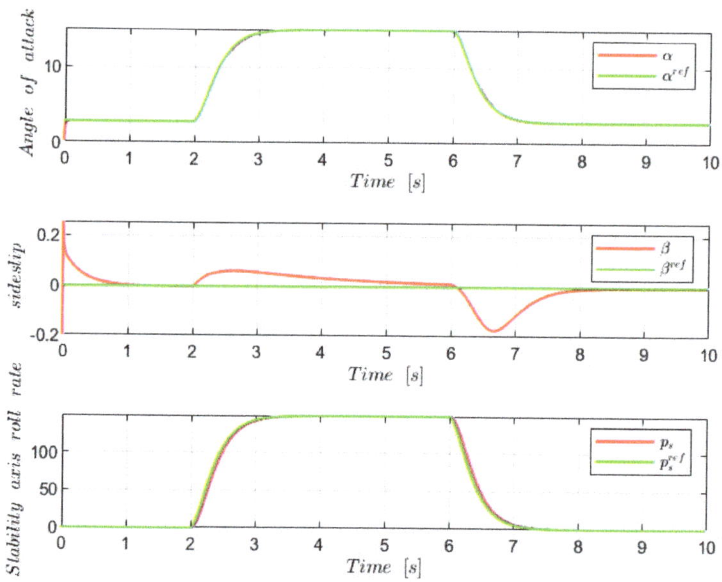

Figure 3. Simulated aircraft control objectives.

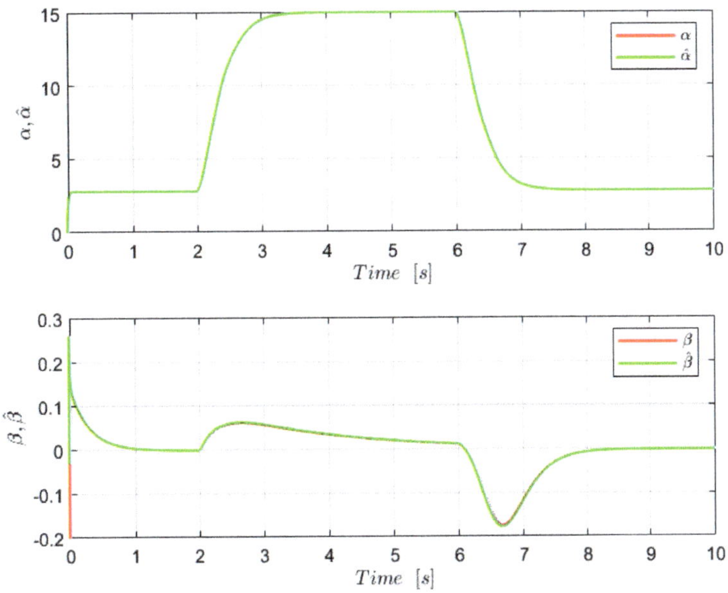

Figure 4. State estimation.

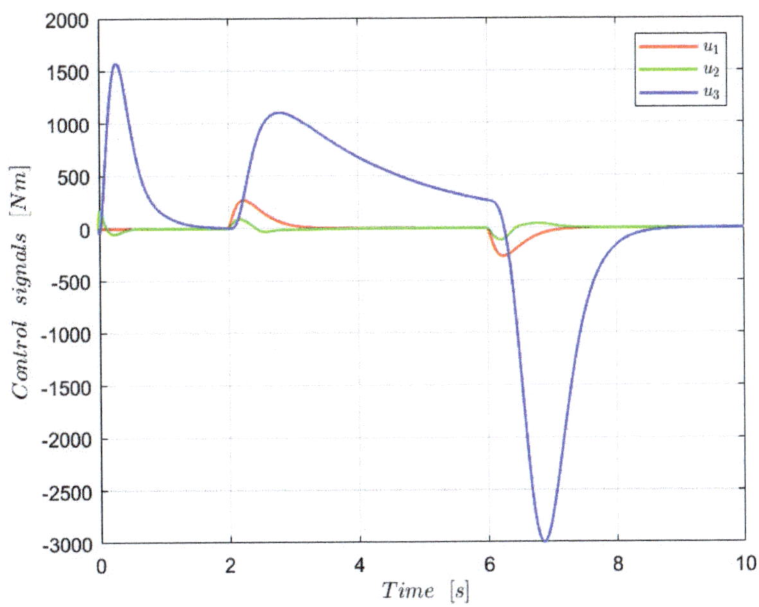

Figure 5. Control signals.

5. Conclusions

In this study, a nonlinear observer backstepping design technique for fighter aircraft control is presented in order to achieve control objectives in the presence of nonlinear lift and side forces. According to the Lyapunov stability theorem, the resultant closed-loop system is exponentially stable for trajectory tracking errors. The simulation results of Admire model seemed to be excellent using this design technique. Furthermore, the observer perfectly estimates the state vector.

Author Contributions: Conceptualization, B.M.S.; methodology, B.M.S.; software, B.M.S. and B.M.; writing—original draft preparation, B.M.S.; writing—review and editing, B.M.S., B.M. and B.A. All authors have read and agreed to the published version of the manuscript.

Funding: This work was supported by the Directorate-General for Scientific Research and Technological Development/Ministry of High Education and Scientific Research of Algeria (DGRSDT/MESRS).

Institutional Review Board Statement: Not applicable.

Informed Consent Statement: Not applicable.

Data Availability Statement: Not applicable.

Conflicts of Interest: The authors declare no conflict of interest.

Abbreviations

The following abbreviations are used in this manuscript:

FOC	Field-Oriented Control
OIB	Observed-Integrator Backstepping
NDM	Nonlinear Damping-Matched
GAM	Generic Aerodata Mode

References

1. Kokotovic, P.V. The joy of feedback: Nonlinear and adaptive. *IEEE Control Syst. Mag.* **1992**, *12*, 7–17.
2. Kanellakopoulos, I.; Kokotovic, P.; Morse, A. A toolkit for nonlinear feedback design. *Syst. Control Lett.* **1992**, *18*, 83–92. [CrossRef]
3. Harrouz, A.; Becheri, H.; Colak, I.; Kayisli, K. Backstepping control of a separately excited DC motor. *Electr. Eng.* **2018**, *100*, 1393–1403. [CrossRef]
4. Abdulgalil, F.; Siguerdidjane, H. Backstepping design for controlling rotary drilling system. In Proceedings of the 2005 IEEE Conference on Control Applications, CCA 2005, Toronto, ON, Canada, 28–31 August 2005; pp. 120–124.
5. Hu, J.; Dawson, D.; Qian, Y. Position tracking control for robot manipulators driven by induction motors without flux measurements. *IEEE Trans. Robot. Autom.* **1996**, *12*, 419–438.
6. Hu, J.; Dawson, D.; Anderson, K. Position control of a brushless DC motor without velocity measurements. *IEE Proc.-Electr. Power Appl.* **1995**, *142*, 113–122. [CrossRef]
7. Xiong, P.; Sun, D. Backstepping-based DPC strategy of a wind turbine-driven DFIG under normal and harmonic grid voltage. *IEEE Trans. Power Electron.* **2015**, *31*, 4216–4225. [CrossRef]
8. Nadour, M.; Essaki, A.; Nasser, T. Comparative analysis between PI & backstepping control strategies of DFIG driven by wind turbine. *Int. J. Renew. Energy Res.* **2017**, *7*, 1307–1316.
9. Bossoufi, B.; Karim, M.; Lagrioui, A.; Taoussi, M.; Derouich, A. Observer backstepping control of DFIG-Generators for wind turbines variable-speed: FPGA-based implementation. *Renew. Energy* **2015**, *81*, 903–917. [CrossRef]
10. Kahveci, N.E.; Ioannou, P.A. Adaptive steering control for uncertain ship dynamics and stability analysis. *Automatica* **2013**, *49*, 685–697. [CrossRef]
11. Strand, J.P.; Ezal, K.; Fossen, T.I.; Kokotović, P.V. Nonlinear control of ships: A locally optimal design. *IFAC Proc. Vol.* **1998**, *31*, 705–710. [CrossRef]
12. Binh, N.T.; Tung, N.A.; Nam, D.P.; Quang, N.H. An adaptive backstepping trajectory tracking control of a tractor trailer wheeled mobile robot. *Int. J. Control Autom. Syst.* **2019**, *17*, 465–473. [CrossRef]
13. Truong, T.N.; Vo, A.T.; Kang, H.-J. A backstepping global fast terminal sliding mode control for trajectory tracking control of industrial robotic manipulators. *IEEE Access* **2021**, *9*, 31921–31931. [CrossRef]
14. Belkheiri, M.; Boudjema, F. Backstepping control augmented by neural networks for robot manipulators. In Proceedings of the AIP Conference Proceedings of First Mediterranean Conference on Intelligent Systems and Automation, American Institute of Physics, Annaba, Algeria, 30 June–2 July 2008; pp. 115–119.

15. Belkheiri, M.; Boudjema, F. Neural network augmented backstepping control for an induction machine. *Int. J. Model. Identif. Control* **2008**, *5*, 288–296.
16. Labbadi, M.; Cherkaoui, M. Robust adaptive backstepping fast terminal sliding mode controller for uncertain quadrotor UAV. *Aerosp. Sci. Technol.* **2019**, *93*, 105306. [CrossRef]
17. Su, Z.; Wang, H.; Yao, P.; Huang, Y.; Qin, Y. Back-stepping based anti-disturbance flight controller with preview methodology for autonomous aerial refueling. *Aerosp. Sci. Technol.* **2017**, *61*, 95–108. [CrossRef]
18. Singh, S.N.; Chandler, P.; Schumacher, C.; Banda, S.S.; Pachter, M. Nonlinear adaptive close formation control of unmanned aerial vehicles. *Dyn. Control* **2000**, *10*, 179–194. [CrossRef]
19. Singh, S.N.; Steinberg, M. Adaptive control of feedback linearizable nonlinear systems with application to flight control. *J. Guid. Control Dyn.* **1996**, *19*, 871–877. [CrossRef]
20. 20. Steinberg, M.L.; Page, A.B. Nonlinear adaptive flight control with genetic algorithm design optimization. *Int. J. Robust Nonlinear Control* **1999**, *9*, 1097–1115. [CrossRef]
21. Krstic, M.; Kokotovic, P.V.; Kanellakopoulos, I. *Nonlinear and Adaptive Control Design*, 1st ed.; John Wiley & Sons, Inc.: New York, NY, USA, 1 October 1995.
22. Herbst, W.B. Future fighter technologies. *J. Aircr.* **1980**, *17*, 561–566. [CrossRef]
23. Well, K.; Faber, B.; Berger, E. Optimization of tactical aircraft maneuvers utilizing high angles of attack. *J. Guid. Control Dyn.* **1982**, *5*, 131–137. [CrossRef]
24. Stevens, B.L.; Johnson, E.N.; Lewis, F.L. *Aircraft Control and Simulation*; Wiley: New York, NY, USA, 1992.
25. Boiffier, J.-L. *The Dynamics of Flight, The Equations*; Wiley: New York, NY, USA, 1998.
26. Härkegård, O.; Glad, S.T. Flight control design using backstepping. *IFAC Proc. Vol.* **2001**, *34*, 283–288. [CrossRef]
27. Marino, R.; Tomei, P. *Global Adaptive Observers and Output-Feedback Stabilization for a Class of Nonlinear Systems*; Springer: Berlin/Heidelberg, Germany, 1991; pp. 455–493.
28. Backström, H. *Report on the usage of the Generic Aerodata Model*; Saab Aircraft AB: Linkoping, Sweden, 1997.

engineering proceedings

MDPI

Proceeding Paper

Pricing Policy Applied to Power Quality Enhancement in Smart Metering Systems †

Abdelmadjid Recioui * and Fatma Zohra Dekhandji

Laboratory of Signals and Systems, Institute of Electrical and Electronic Engineering, University M'hamed Bougara of Boumerdes, Boumerdes 35000, Algeria
* Correspondence: a_recioui@univ-boumerdes.dz

† Presented at the 2nd International Conference on Computational Engineering and Intelligent Systems, Online, 18–20 November 2022.

Abstract: Power quality problems exist in every power system; the more advanced the system is, the higher the probability for issues to occur is, and the smart grid is no exception. In this paper, a study of a residential area is implemented using the LABVIEW simulation software, in which the power of each house with its own appliances is monitored in real time, meaning the power of the whole system is also monitored, including instantaneous and accumulative power, current, power factor—all elements required for system assessment. In addition, a billing policy of each individual house to see the total price to be paid for the power supplied is also devised. All these factors contribute to the analysis and overview of the effect of power quality problems on the grid, both electrically and economically. The results of the simulation indicate that the pricing policy proposed is indeed effective for both sides of the grid, the supplier and the consumer.

Keywords: power quality; smart metering; pricing policy; smart grids

Citation: Recioui, A.; Dekhandji, F.Z. Pricing Policy Applied to Power Quality Enhancement in Smart Metering Systems. *Eng. Proc.* **2023**, *29*, 15. https://doi.org/10.3390/engproc2023029015

Academic Editor: Hamid Bentarzi

Published: 18 January 2023

1. Introduction

Power quality (PQ) is defined as supplying the clearest sinusoidal waveforms for voltage and current for the electrical network at the recommended frequency and magnitude [1–5]. PQ can be influenced by a variety of factors, including voltage and frequency fluctuations, imbalance, interruption, flicker, and harmonics [6–8]. The surveillance and regulating of PQ is a significant hurdle for electricity companies. As a result, energy suppliers do not quantify or sustain PQ for residential customers.

Introducing a PQ-based penalty/incentive strategy for residential clients necessitates the installation of smart meters for each client. Traditional digitizedor smart electricity meters are created to monitor particular electric indicators, such as power, energy, and power factor, among others. Due to hardware and software restrictions, they are not suitable for the deployment of a PQ-based tariffingsystem. PQ meters should also possess extra functionalities, such as higher sampling rates for voltage and current, synchronized sampling, denoising, data processing, and the ability to handle largedata [9].

Engineers, electricians, and technicians primarily use commonly produced PQ analysis tools to track and troubleshoot electronic systems or locations [10]. Yet, these are prohibitively expensive and impractical for an electricity company to install at each residential client's home for direct PQ meter reading. Furthermore, it is difficult for customers to interpret such complicated meters and understand the quality of the electricity they obtain. As a result, an easy-to-usesmart meter including useful energy and PQ monitoringcapabilities togetheris required. Moreover, the meter must be robust, user-friendly, simple to comprehend, and inexpensive.

With a suitable PQ surveillance equipment and the implementation of a PQ-based tariffing for low voltage clients, a whole distribution network can be controlled from the PQ point of view [9]. The penalty or reward pricing scheme based on PQ thresholds must be launched. Based on the techno-economic characteristics of poor PQ, the penalty and incentive value for reactive power level and harmonic content may be determined [11].

Numerous techniques to create a tariffing related to the PQ level have been suggested in the literature. Several countries, including Belgium, the United States, the United Kingdom, Germany, and Poland, have put in place reactive energy tariffs to minimize losses in their distribution networks [12,13]. Ref. [14] suggests a tariff strategy for residential clients based on reactive energy and provides the supplier's savings as well. This strategy, on the other hand, mainlyemphasizes capital expenditure (capex) payback, with price breakdown values estimated experimentally. Other tariff strategies [15,16] provide reactive energy rewards depending on the power factor, active power, or reactive energy, with all reward parameters derived experimentally.

A lot of important work has been conducted lately in the area of PQ-based smart metering. Ref. [17] suggests design criteria and important attributes for a PQ-based smart meter. One more work, [18], illustrates a smart PQ meter for residential clientsthat monitors PQ metrics, voltage and frequency incidents. It does not, however, indicate or track the client's current. Furthermore, it necessitates an outside computer and does not include an energy metering feature. The authors of [19] presented an ethernet-based smart meter for PQ monitoring that assesses voltage quality using the LABVIEW software. It encompasses aload scheduling feature for onlinePQ enhancement. Such ameter further necessitates a separate computer to operate LABVIEW and does not monitor clients' current anomalies. Under solelylinear load situations, a stand-alone smart meter presented in [20] can identify voltage sag and swell, supply frequency, and power factor. Nevertheless, it is not suitable for quantifying the harmonic content for both voltage and current of power–electronic fedloads, neither it is adequate for measuring the displacement power factor (DPF). The authors of [21] created a smart meter relying on a Raspberry Pi that cannot quantify clients' current disturbance. The voltage metrics are the only ones measured by an accessible Raspberry-Pi-based PQ monitoring system described in [22]. Furthermore, the use of Raspberry Pi is repetitive, costly, and has a comparatively greater energy usage. The multi-purpose smart meter in [23] can measure both power and PQ disturbancesand has anti-theft functionalities. However, it missesthe feature ofassessing clients' current distorsion. The authors in [24] propose an advanced FPGA-based PQ surveillance system that turns out to becomplex and requires a peripheral computer to run LABVIEW. One more FPGA-based intelligent power meter described in [25] is a stand-alone instrument for residential applications. However, it turns out to have acomplicated structure that is not cost-effective for residential use.It may be deduced that the smart meters built around FPGA and RaspberryPiare not self-contained and cost-effective for residential PQ surveillance applications.

The aim of this work is to examine the power quality disturbances in areas equipped with smart meters and their effect on the power consumption and pricing from both the consumer's and supplier's perspectives. To fulfill the goals of this study, a simulation of a residential area with a composition of 60 houses equipped with 8 appliances, with houses influenced by the earlier cited power quality problems, has been developed to clarify and detect the slight change in the electrical system parameters, which revealed that power quality problems increase the power demand and the pricing bills for the consumer and the supplier.

2. State of Art on Electrical Energy Pricing Policies

Tariffs are the costsof providing electric power to an end user [26–30]. The quantity of power that the load receives determines the price of generating electricity. Tariffs must be adapted to different kinds of clients (e.g., industrial, residential and commercial). Pricing estimation for variousclients is far more difficult. The pricing policy includes several objectives to be accomplished.

Utility companies sell electricity at a rate that not only encloses its expensesbut generates a reasonable profit as well. Tariff rates should aim to accomplish the following goals:

The expense of generating electrical power at the power station is covered. Costs associated with the capital spending in distribution and transmission infrastructure. Restoration of the operating and maintenance expenses for electrical power availability. A reasonable return on capital funding.There are several types of tariffing strategies [30–33]. These are outlined in what follows:

2.1. Simple Tariff

A simple pricing or standard price has a specified priceper amount of electrical power expended. This tariff scheme is straight forward, and the cost does not vary with the number of units used. An energy meter is used to track the amount of electricity spent at the client connectors. Clients can understand this tariffing scheme because it is rudimental. Every consumer should reimburse the same set price irrespective of usagevariability. The price per supplied amountis substantial. It does not advocate for the consumption of energy.

2.2. Flat Rate Tariff

A flat rate tariff refers to when various kinds of clients are billed at diverse uniform per unit prices. In this scheme, clients are divided into various categories, and each category has a specified uniform price. The various consumer categories are created with their variability and load conditions in mind. This scheme is more beneficial to the various kinds of clients. Flat rate tariffing is very easy in terms of calculations. Distinct meters are needed for lighting load, power load, and so on. The implementation of such a tariff is complex and expensive. A specific category of clients is billed at the same price regardless of the amount of electricity utilized.

2.3. Block Rate Tariff

Block rate tariffing means a particular interval of electricity is required to be paid at a specified priceand subsequent levels of electricity are billed at gradually lower prices [30]. The electricity usage is grouped into clusters, with a set cost per share at every cluster. The cost per electricity usagein the first cluster is the most expensive and it is gradually decreased for the subsequent intervals of consumption. The client receives a reward for utilizing more electricity. This enhances the power factor of the network and, thus, the electricity generation expenses are lowered. Nevertheless, this scheme fails to monitor the clients' electricity demand. This kind of tariffing is deployed in most domestic and low-consumption commercial clients.

2.4. Two-Part Tariff

A two-part tariffing refers to the case when the price of electricity is billed relative to the user's peak power as well as the utilized amounts of energy [30–33]. The overall client's billis separated into two parts in a two-part tariffing scheme: fixed costs and running fees. The fixed costs are determined relative tothe user's peak power, whereas the running fees are measured in terms of power consumed. As a result, the client is billed a set fee per kW of peak power plus a predefined costper kWh of electricity expended. This kind of tariffingis typically utilized with industrial customers that havehigher peak power. Users can easily understand this scheme. It restores the expenses that are derived fromthe users' peak power. This scheme is unaffected by the amount of energy spent. The client should reimburse the direct costs regardless of the amount of electricity utilized. There arefrequently mistakes in estimating the peak user power usage.

2.5. Maximum Demand Tariff

This is similar to a two-part tariffing scheme; however, the peak power is actually monitored by installing a peak loaddemand meter in the user's home. The highest usage is determined solely by the presale estimate. This tariffing strategyis often applied to large customers. This is not appropriate for relativelysmallerclients. It is necessary to have a dedicated maximum power meter.

2.6. Power Factor Tariff

The power factor tariffingis the pricethat considers the user's load power factor. A poor power factor raises the rated capacity of the powerstation transformers and multiplies transmission losses. A user with a poor power factor must be punished. The nextare some examples of power factor tariffs: kVA pricefor peak consumption: It is a two-part tariffing systemin a mutated way. In this situation, the set costs are calculated based on the highest consumption in kVA rather than kW. A client with a poor power factor must make a significant contribution to the fixed costs. Tariffing based on a relative scale: This is additionally namedan average power factor tariffing.Within that scenario, a mean power factor, assume 0.8 lagging, is used as a guide. In case the user's power factor drops below this level, appropriate extra fees are levied. If the power factor is greater than a certain threshold, the client is entitled to a price reduction.

2.7. KW and KVAR Tariff

This tariffing formcharges for both active power (kW) and reactive power (kVAR) delivered [34,35]. A client with a lower power will use more reactive power and thus reimburse more expenses.

2.8. Three-Part Tariff

A three-part tariffing implies thatthe total cost to bepaid by the client is divided into three parts: a fixed fee, a semi-fixed cost, and an operating price, and is given as [30]:

$$\text{Total charge} = \text{Rs}(a + b \times \text{kW} + c \times \text{kWh})$$

where:

a = setfee made at each bill establishement phase;

b = cost per kW of peakpower usage;

c = price per kWh of electricity utilized.

It is evident that appending a fee to a two-part tariffing schemeresults in a three-part tariff. The main criticism of this form of tariffingis that the costsare divided into three parts. This strategyis typically imposed on large consumingclients.

3. Results and Discussions

3.1. Simulated System Description

Using LABVIEW software, a simulation of a smart city with 60 houses numbered from 1 to 60 is performed in order to investigate and quantify some major power quality problems. The city has been arbitrarily divided into sectors, each sector with a number of houses, and each house with different appliances. The six sectors andtheir corresponding houses are: houses 1–12 for sector 1, 13–20 for sector 2, 21–33 for sector 3, 34–40 for sector 4, 41–51 for sector 5, and 52–60 for the last sector. Each house is equipped with different appliances: TV, washer, dryer, iron, fridge, microwave, computer and heater.Voltage and current waveforms are generated with amplitude and phase calculated from power factor consumption, withthe voltage as a fixed value of 220 V, a phase of 0 rad, and a frequency of 50 Hz. As for the current, it is deduced from the power consumption factor and the power factor for each appliance. Table 1 presents the real life estimated usage power for each appliance to make the simulation as close as possible to real life scenarios.

Table 1. List of appliances' estimated wattage.

Common Appliances	Estimated Wattage
TV 42" Plasma	240
Washing Machine	500–1000
Hair Dryer	1000–1875
Clothes Iron	1000–1500
Refrigerator	500–750
Microwave Oven	1000–2000
Computer Laptop	20–75
Space Heater	750/1500

To account for harmonic content, harmonics are added to the sinusoidal waveform. Three block diagrams of signals have been added with frequency multiple of the fundamental frequency to imitate the effect of the third, fifth, and seventh harmonics. The calculation of the total harmonic distortion (THD) is carried outby taking the ratio of the square root of the sum of the square RMS values of the current harmonic signals to the RMS value of the fundamental current signal.

3.2. Simulation at One House Level

Figure 1 shows the appliances' ON/OFF switches grouped in one cluster, the electricity bill that demonstrates the different fees, the price without fees and the total price, andthe house parameters outcomes such ascurrent, instantaneous power, accumulative power, apparent power, real power and power factor. Figure 2 presents the detailed block diagram for that house.In the block diagram, eight sub-VIs that simulate the eight house appliances, each with its own ON/OFF switch button are presented. The KWh consumed instantaneously and cumulatively, the current, the apparent power and real power of the all eight appliances outputs are added to obtain the total instantaneous power, the total cumulative power, the overall current, the total real power and the total apparent power. Then, the resultant total real power is divided over the total apparent power in order to obtain the power factor.

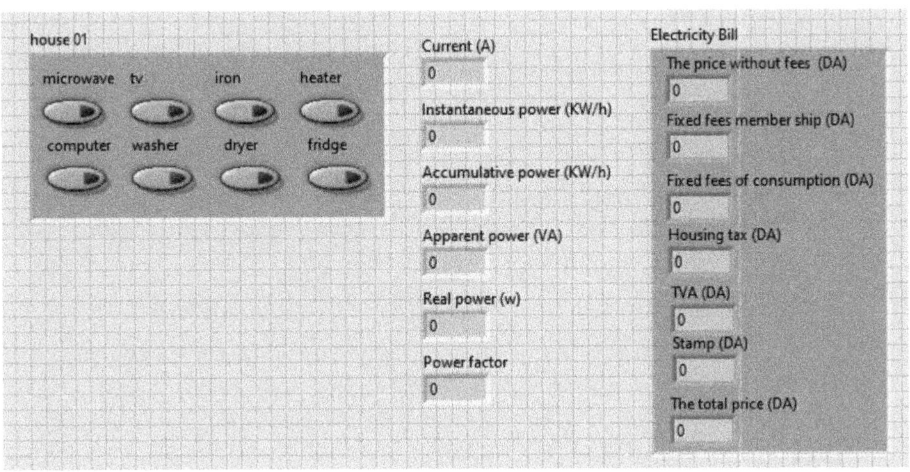

Figure 1. Front panel of house 01.

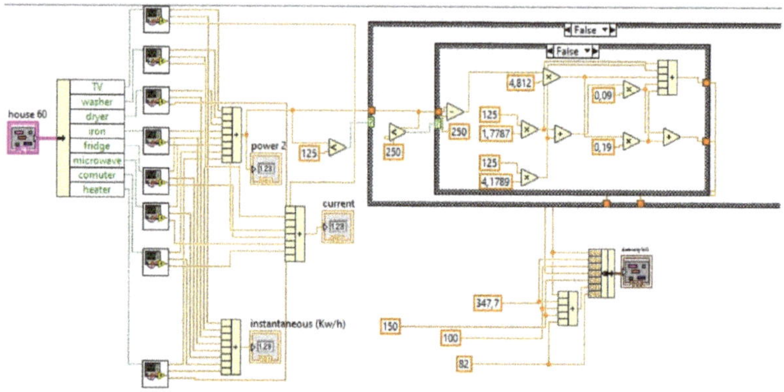

Figure 2. House 01 block diagram.

3.3. Proposed Pricing Policy for Power Quality Improvement

A penalty/incentive pricing policy is deployed with the objective of encouraging electricity consumers to provide devices that correct and eliminate the harmful power quality issues, along with shaping their consumption manners is such a way that respects the criteria of healthy electric grid. Four power quality issues are considered in this incentive/penalty policy. For the power factor, a poor power factor (below 90%) results in a 20% penalty (of the bill price) as an additional cost to be paid by the client, while no extra fees are applied if this power factor is within the limit (>90%). For the THD, if it is below 5%, a 20% incentive (deduced from the overall bill) applies. If the THD is more than 5% but less than 10%, a 30% penalty is charged for the user. If the THD exceeds 10%, a 50% penalty is applied. For the voltage sags and swells, a ±20% limit for a 220 V nominal value should not be exceeded. If the voltage falls below 209 or exceeds 231, a 50% penalty is applied, while no extra fees are charged if the voltage is confined in the interval [209–231]V.

3.4. Results and Discussion

Figure 3 represents the front panel of the whole smart city that is composed of six sectors where each sector provided with two clusters: one that contains current and accumulative power numeric indicators of each house that is a part of the sector, and a second one that contains all the ON/OFF buttons of appliances of each house in the sector. There are three graphs for each sector with digital visual indicators: the total accumulative power consumption, the total instantaneous power consumption, and the total current of the sector. A numerical indicator provides the amount of money that the consumer is paying for the electrical company supplier for each sector and for the entire smart city, in addition to a graph with digital indicator of the sum of currents and instantaneous consumption power of all the six sectors.

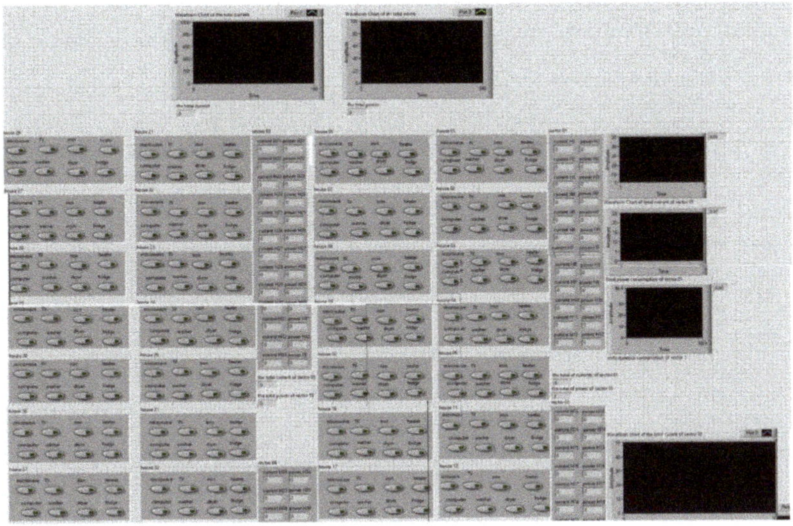

Figure 3. Front panel of the simulated smart city.

Figure 4 represents the front panel of the smart city's electrical grid that provides the cumulative and instantaneous power consumption, total current, total incentive prices, total penalties prices, total price with penalty/incentive, and total price without penalty/incentive. These are presented in both numerical indicators and using graphical representation. It is noticed that, duringthe peak hours, the total current reaches its largest values, which indicates that the demand on power has peaked due to the increase in the power consumption of the houses. At the end of the simulation, it has been observed that the amount of cumulative power and the total price without incentive/penalty were considerable. This can be explained by the effect of power quality issues that raised the voltage and the current, leading the power consumption to be increased, which means that the supplier is required to provide more power than the normal amount in order to satisfy the demand. It can be noticed that deploying the incentive and penalty pricing policy results in more money earned by the supplying company, which can be used to cover the additional expenses of the additional power and entice the faulty consumers to correct their PQ issues.

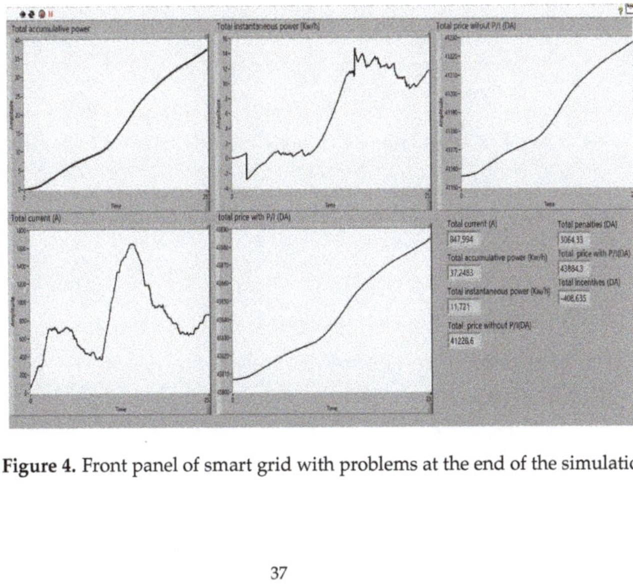

Figure 4. Front panel of smart grid with problems at the end of the simulation.

Figure 5 presents the front panel of the smart grid after the correction of the detected PQ problems. The total penalties are insignificant and the incentives are substantial, indicating that a large number of users have fixed their PQ problems. The peak power demand and the final accumulative power have been lowered compared to the values of the system with the power quality problems. This reduced the amount of additional power that is consumed with the existing PQ problems, resulting in a decrease in the utility company investments to produce more electricity. It is remarked that a slight difference exists between the total price with and without an incentive/penalty policy, which justifies its usefulness, as there is additional financial stress on the utility company.

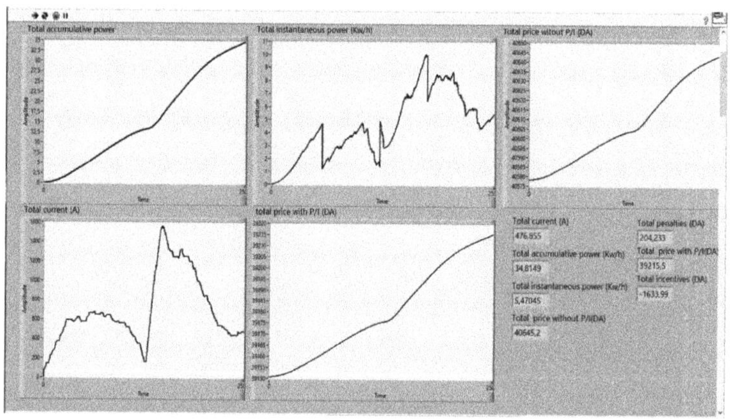

Figure 5. Front panel of smart grid after PQ problems correction at the end of simulation.

4. Conclusions

In this paper, a simulation of a smart area with a mix of houses are affected by four kinds of power quality problems (voltage swell, voltage sag, poor power factor and harmonics) has been presented, along with some healthy houses. A pricing policy has been employed to cut down the effect of these problems by encouraging and raising awareness of consumers to adopt a healthy consumption behavior. The simulation results strengthened the basic idea that the pricing policy used by the supplier companies is not practical in the process of fixing power quality problems because it provides a strategy that urges clients to correct their PQ problems. The ordinary billing counts the excess of power resulting from PQ issues as an ordinary power with the same price as the consumed one. This policy has bad financial effects on both the utility company, which invests more money with the aim of generating excess power to satisfy the power demand, and the client, who pays for extra, unnecessary power.

Author Contributions: Conceptualization, F.Z.D. and A.R.; methodology, A.R.; software, F.Z.D.; validation, A.R.; formal analysis, F.Z.D.; investigation, A.R.; resources, F.Z.D.; data curation, F.Z.D.; writing—original draft preparation, F.Z.D.; writing—review and editing, A.R.; visualization, F.Z.D.; supervision, A.R.; project administration, A.R. All authors have read and agreed to the published version of the manuscript.

Funding: No external funding was received to carry out this research.

Institutional Review Board Statement: Not applicable.

Informed Consent Statement: Not applicable.

Data Availability Statement: The data presented in this study are available on request from the corresponding author. The data have been synthesized for the purpose of this study and are not publicly available.

Conflicts of Interest: No conflicts of interest are declared by the authors.

References

1. Dekhandji, F.Z.; Talhaoui, S.; Arkab, Y. Power Quality Detection, Classification and Monitoring Using LABVIEW. *Alger. J. Signals Syst.* **2019**, *4*, 101–111. [CrossRef]
2. Dekhandji, F.Z.; Rais, M.C. A Comparative Study of Power Quality Monitoring Using Various Techniques. In *Optimizing and Measuring Smart Grid Operation and Control*; IGI Global: Hershey, PA, USA, 2021; pp. 259–288.
3. Dekhandji, F.Z. Detection and mitigation of power quality problems based on phasor measurment units and renewable energy systems. In Proceedings of the 3rd International Conference on Computational and Experimental Science and Engineering (ICCESEN 2016), Antalya, Turkey, 19–24 October 2016.
4. Dekhandji, F.Z.; Douche, M.; Zebidi, N. DVR and D-STATCOM Mitigation Techniques of Power Quality Effects on Induction Motors. *Alger. J. Signals Syst.* **2017**, *2*, 110–129. [CrossRef]
5. Chattopadhyay, S.; Mitra, M.; Sengupta, S. Electric power quality. In *Electric Power Quality*; Chattopadhyay, S., Mitra, M., Sengupta, S., Eds.; Springer: Dordrecht, The Netherlands, 2011; pp. 5–12.
6. Kappagantu, R.; Daniel, S.A.; Yadav, A. Power quality analysis of smart grid pilot project, Puducherry. *Procedia Technol.* **2015**, *21*, 560–568. [CrossRef]
7. Moses, P.S.; Deilami, S.; Masoum, A.S.; Masoum, M.A. Power quality of smart grids with plug-in electric vehicles considering battery charging profile. In Proceedings of the IEEE PES Innovative Smart Grid Technologies Conference Europe (ISGT Europe), Gothenberg, Sweden, 11–13 October 2010; pp. 1–7.
8. Niitsoo, J.; Jarkovoi, M.; Taklaja, P.; Klüss, J.; Palu, I. Power quality issues concerning photovoltaic generation and electrical vehicle loads in distribution grids. Smart Grid Renew. *Energy* **2015**, *6*, 164.
9. Balwani, M.R.; Thirumala, K.; Mohan, V.; Bu, S.; Thomas, M.S. Development of a Smart Meter for Power Quality-Based Tariff Implementation in a Smart Grid. *Energies* **2021**, *14*, 6171. [CrossRef]
10. Viciana, E.; Alcayde, A.; Gil Montoya, F.; Baños, R.; Arrabal-Campos, F.M.; Zapata-Sierra, A.; Manzano-Agugliaro, F. OpenZmeter: An Efficient Low-Cost Energy Smart Meter and Power Quality Analyzer. *Sustainability* **2018**, *10*, 4038. [CrossRef]
11. Da Silva, R.P.B.; Quadros, R.; Shaker, H.R.; da Silva, L.C.P. Effects of mixed electronic loads on the electrical energy systems considering different loading conditions with focus on power quality and billing issues. *Appl. Energy.* **2020**, *277*, 115558. [CrossRef]
12. ACER European Union Agency for the Cooperation of Energy Regulatory. ACER Report on Distribution Tariff Methodologies in Europe. February 2021. Available online: https://documents.acer.europa.eu/Official_documents/Acts_of_the_Agency/Publication/ACER%20Report%20on%20D-Tariff%20Methodologies.pdf (accessed on 1 June 2022).
13. Bućko, P.; Wilczyński, A. Reactive Energy Billing in Operator Tariffs in Poland. *Acta Energetica* **2015**, *4*, 4–9. [CrossRef]
14. Ahmad, A.; Kashif, S.A.R.; Saqib, M.A.; Ashraf, A.; Shami, U.T. Tariff for reactive energy consumption in household appliances. *Energy* **2019**, *186*, 115818. [CrossRef]
15. Jay, D.; Swarup, K.S. Game theoretical approach to novel reactive power ancillary service market mechanism. *IEEE Trans. Power Syst.* **2020**, *36*, 1298–1308. [CrossRef]
16. Cerbantes, M.C.; Fernandez-Blanco, R.; Ortega-Vazquez, M.A.; Mantovani, J.R.S. Incorporating a Nodal Reactive Power Pricing Scheme Into the DisCo's Short-Term Operation. *IEEE Trans. Smart Grid* **2018**, *10*, 3720–3731. [CrossRef]
17. Chang, T.Y.; Wang, Y.; Sun, H.J. The Design of Smart Meter with Power Quality Monitoring. *Electr. Meas. Instrum.* **2012**, *49*, 74–77.
18. Ramos, N.R.; Pereira, P.; Martins, J. Smart-meter in power quality. In Proceedings of the 2017 International Young Engineers Forum (YEF-ECE), Costa da Caparica, Portugal, 5 May 2017; Institute of Electrical and Electronics Engineers (IEEE): New York, NY, USA, 2017; pp. 42–46.
19. Das, H.; Saikia, L. Ethernet based smart energy meter for power quality monitoring and enhancement. In Proceedings of the 2017 Recent Developments in Control, Automation & Power Engineering (RDCAPE), Noida, Indi, 26–27 October 2017; Institute of Electrical and Electronics Engineers (IEEE): New York, NY, USA, 2017; pp. 187–191.
20. Madhu, G.M.; Vyjayanthi, C.; Modi, C.N. Design and Development of a Novel IoT based Smart Meter for Power Quality Monitoring in Smart Grid Infrastructure. In Proceedings of the TENCON 2019 IEEE Region 10 Conference (TENCON), Kochi, India, 17–20 October 2019; pp. 2204–2209.
21. Castello, P.; Muscas, C.; Pegoraro, P.A.; Sulis, S. Low-Cost Energy Meter with Power Quality Functionalities. In Proceedings of the 24th Imeko TC, 2020; Available online: https://www.imeko.org/publications/tc4-2020/IMEKO-TC4-2020-56.pdf (accessed on 1 June 2022).
22. Medvedr, D.; Conka, Z.; Kolcun, M.; Kanalik, M.; Zbojovsky, J.; Pavlik, M. Project design of the electric power quality analyzer using an open-source platform. In Proceedings of the 2018 International IEEE Conference and Workshop in Óbuda on Electrical and Power Engineering (CANDO-EPE), Budapest, Hungary, 20–21 November 2018; pp. 000169–000172.
23. Amankhan, A.; Kural, A.; Temirbek, I.; Abukhan, A.; Mukashov, D.; Azamat, A.; Kudaibergenov, K.; Bagheri, M. Multi-functional Smart Electricity Metering System. In Proceedings of the 2019 IEEE International Conference on Environment and Electrical Engineering and 2019 IEEE Industrial and Commercial Power Systems Europe (EEEIC/I&CPS Europe), Genova, Italy, 11–14 June 2019; pp. 1–6.

24. Xavier, M.V.E.; Boaventura, W.D.C.; Noronha, M.D. High performance power quality monitoring system. In Proceedings of the 2016 17th International Conference on Harmonics and Quality of Power (ICHQP), Belo Horizonte, Brazil, 16–19 October 2016; pp. 974–979.
25. Arenas, L.D.O.; Melo, G.D.A.E.; Canesin, C.A. A Methodology for Power Quantities Calculation Applied to an FPGA-Based Smart-Energy Meter. *IEEE Trans. Instrum. Meas.* **2021**, *70*, 1–11. [CrossRef]
26. Dekhandji, F.Z. Signal Processing Deployment in Power Quality Disturbance Detection and Classification. *Acta Phys. Pol. A* **2017**, *132*, 415–419. [CrossRef]
27. Arrillaga, J.; Watson, N.R.; Chen, S. *Power System Quality Assessment. 1.2 Disturbances*; Wiley-Blackwell: Hoboken, NJ, USA, 2000.
28. De Almeida, A.; Moreira, L.; Delgado, J. Power quality problems and new solutions. *Int. Conf. Renew. Energies Power Quality* **2003**, *3*, 1–45. [CrossRef]
29. Bingham, R.P. Sags and Swells. Original Draft September 1994. Revised 16 February 1998. Available online: http://www.northwoodpower.ca/technicaldocuments/sags-and-swells.pdf (accessed on 1 June 2022).
30. Unit V: Energy, Economic and Environmental Issues of Power Plants. Available online: https://stannescet.ac.in/cms/staff/qbank/EEE/Notes (accessed on 1 June 2022).
31. Price, M.; Allmeroth, T.; Cleveland, M.; Regenwether, J. *Implementing Dynamic Pricing Meter Configuration Trade-Offs*; Deloitte Development LLC: Tirana, Albania, 2012.
32. Dekhandji, F.Z.; Recioui, A. An Investigation into Pricing Policies in Smart Grids. *Eng. Proc.* **2022**, *14*, 15. [CrossRef]
33. Dassa, K.; Recioui, A. Application of Optimization to Sizing Renewable Energy Systems and Energy Management in Microgrids: State of the Art and Trends. In *Applications of Nature-Inspired Computing in Renewable Energy Systems*; IGI Global: Hershey, PA, USA, 2022; pp. 60–94.
34. Meehan, P.; McArdle, C.; Daniels, S. An Efficient, Scalable Time-Frequency Method for Tracking Energy Usage of Domestic Appliances Using a Two-Step Classification Algorithm. *Energies* **2014**, *7*, 7041–7066. [CrossRef]
35. Mukherjee, S.; Ganguly, A.; Paul, A.K.; Kumar Datta, A. Load Flow Analysis and Reactive Power Compensation. In Proceedings of the 2018 International Conference on Computing, Power and Communication Technologies (GUCON), Greater Noida, India, 28–29 September 2018; pp. 207–211.

engineering proceedings

MDPI

Proceeding Paper

Suspension System Control Process for Buses with In-Wheel Motors †

Mohamed Belrzaeg [1], Abdussalam Ali Ahmed [2,*], Mohamed Mohamed Khaleel [3], Abdulgader Alsharif [4], Maamar Miftah Rahmah [5] and Ahmed Salem Daw Alarga [6]

1 College of Applied Sciences and Technology, Al-Awata, Libya
2 Mechanical and Industrial Engineering Department, Bani Waleed University, Bani Walid, Libya
3 Electrical and Electronic Department, Karabuk University, Karabuk 78050, Turkey
4 School of Electrical Engineering, Universiti Teknologi Malaysia, Skudai 81310, Malaysia
5 Mellitah Oil and Gas Company, Tripoli, Libya
6 Computer Science Department, Elmergib University, Khums, Libya
* Correspondence: abdussalam.a.ahmed@gmail.com
† Presented at the 2nd International Conference on Computational Engineering and Intelligent Systems, Online, 18–20 November 2022.

Abstract: In the last few years, there has been considerable growth in in-wheel electric motors manufacturing and the number of electric buses operating around the world. As a result of this clear increase, competition between electric bus manufacturers is increasing to reach the best satisfaction and comfort for passengers. This paper aims to deliver significantly better results due to suspension system. This paper also aims to evaluate an active bus suspension system with an in-wheel electric motor and to show the effect of this motor on the performance of the bus's suspension system. In this work, a quarter bus suspension system is simulated and modelled using Matlab software, in addition to using one of the well-known control technologies, which is the linear quadratic regulator (LQR). The results showed that the weight of the electric motors in the bus's wheels had a slightly negative effect on passenger comfort, as the reason for this effect is because the electric motor increased the mass of the wheel.

Keywords: electric bus; active suspension system; in-wheel motors; modelling; LQR

Citation: Belrzaeg, M.; Ahmed, A.A.; Khaleel, M.M.; Alsharif, A.; Rahmah, M.M.; Alarga, A.S.D. Suspension System Control Process for Buses with In-Wheel Motors. *Eng. Proc.* **2023**, *29*, 4. https://doi.org/10.3390/engproc2023029004

Academic Editors: Abdelmadjid Recioui, Hamid Bentarzi and Fatma Zohra Dekhandji

Published: 11 January 2023

1. Introduction

The first part of this paper includes simplified information about the growth and increase in the number of electric buses EBs in the world, in addition to the definition and structure of in-wheel electric motors (IWMs). In addition, in this part, the importance of the active suspension system is mentioned, as well as the control method used, which is the LQR control.

Clean public transportation systems are becoming increasingly important in today's cities to combat air pollution. In most of the world's countries, energy and environmental policies have increased recommendations and incentives to introduce a large amount of clean technology for vehicles, including electric buses, by 2030 [1]. In recent years, there has been a significant increase in the use of electric buses around the world, particularly in China.

Electric vehicles have contributed to a more than 3% reduction in the growth of oil use. Moreover, electric buses account for around 3/4 of the reduction in this consumption. As evidence of this, a significant portion of our energy transformation will require the widespread adoption of electrified public transportation.

In 2018, Bloomberg New Energy Finance, one of the key research providers, produced a report titled "Electric Buses in Cities: Driving towards Cleaner Air and Lower CO_2", which forecasted a tripling of the number of electric buses by 2025, as illustrated in Figure 1.

That is 1.2 million buses, or roughly half of all buses currently in service, that would be converted to electric buses [2].

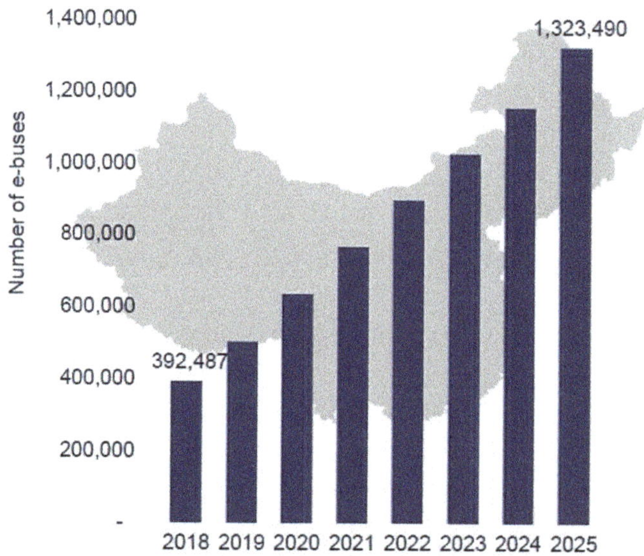

Figure 1. The predicted number of electric buses (EBs) in the world by 2025.

According to Wood Mackenzie Power and Renewables, adoption of electric buses in China will expand dramatically from 2018 to 2025 [3], as illustrated in Figure 2.

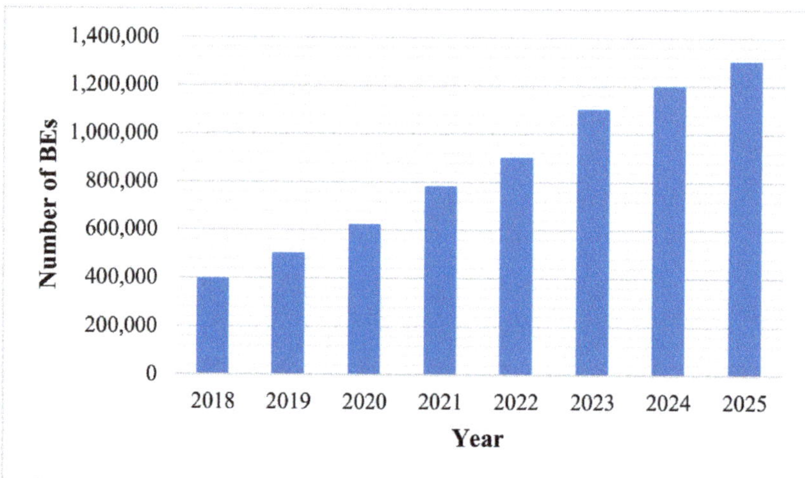

Figure 2. The number of electric buses in China from 2018 to 2025.

One sort of electric vehicle drive system is the in-wheel motor that is shown in Figure 3. In conventional electric vehicles, the gasoline engine is replaced by an electric motor; this motor incorporates motors directly into each of the driving wheels.

Figure 3. IWM structure.

Electric motors are used in the wheels of electric vehicles, and they are called in-wheel electric motors (IWMs), where these motors increase the weight of the wheels clearly, and this increase in weight affects the performance of the suspension system in electric buses.

Suspension is a crucial aspect of a vehicle's design because its purpose is to provide comfort while driving in a variety of road and weather situations. Suspension analyses were carried out in a study in order to analyze a vehicle's performance with an increased mass on each wheel [4].

Numerous control techniques have been offered to address these suspension issues. Ride comfort is improved by reducing the body acceleration induced by the car's body when it encounters road disturbances from both smooth and real road roughness, which is accomplished using a fuzzy control algorithm [5,6]. Numerous active suspension control techniques, including linear quadratic Gaussian control, adaptive control, and nonlinear control, have been developed and proposed to address these issues [7–9]. In this work, the LQR method was chosen to evaluate the active suspension system of an electric bus.

2. Bus Suspension System Model

The control problem of designing any vehicle suspension system is both exciting and demanding. When designing a suspension system, a quarter model (one of the four vehicle wheels) is employed to reduce the problem to a single-dimensional multiple spring–damper system. The following figure is a diagram of this system. This model is for an active suspension system that includes an actuator capable of generating the actuating force (A.F) required to control the bus body's motion and the wheel deflection.

A description and the values of the variables of the active suspension system of the electric bus are shown in both Table 1 and Figure 4.

Table 1. Bus suspension parameters.

Parameter	Value, Unit
Quarter of bus mass (m_b)	2450 kg
Suspension stiffness (S_1)	79,000 N/m
Wheel stiffness (S_2)	500,100 N/m
Wheel mass or unsprung mass (m_w)	330 kg
The damping coefficient of the bus wheel (d_2)	15,000 N.s/m
The damping coefficient of the bus system (d_1)	355 N.s/m

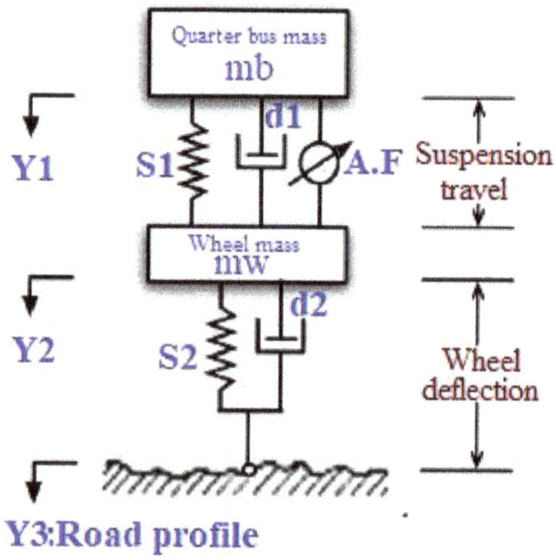

Figure 4. A quarter bus model.

According to the system depicted in Figure 4, the system parameters stated in Table 1, and Newton's law, we can formulate the system's dynamic equations as follows:

$$m_b\ddot{Y}_1 = -S_1(Y_1 - Y_2) - d_1\left(\dot{Y}_1 - \dot{Y}_2\right) + A.F \quad (1)$$

$$m_w\dot{Y}_2 = S_1(Y_1 - Y_2) + d_1\left(\dot{Y}_1 - \dot{Y}_2\right) + A.F - S_2(Y_2 - Y_3) - d_2\left(\dot{Y}_2 - \dot{Y}_3\right) - A.F \quad (2)$$

where:

$A.F$: the actuator force
Y_3: the road surface profile
$Y_1 - Y_2$: represents the suspension travel
$\dot{Y}_1$: denotes the velocity of the bus body or the sprung mass
$\ddot{Y}_1$: symbolizes the acceleration of the bus body
$Y_2 - Y_3$: denotes the wheel deflection
Y_2: symbolizes the vertical velocity of the wheel

The two main parameters that are taken into account when designing suspension systems are the acceleration of the bus body, which is symbolized by "$\ddot{Y}_1$" and the second parameter is the deflection of the wheel, symbolized by "$Y_2 - Y_3$", where the first symbol indicates passenger comfort while the second symbol indicates road holding.

According to all the previous data represented by the system diagram in addition to the two dynamic equations, it is possible to create a Simulink model for the bus as follows (Figure 5):

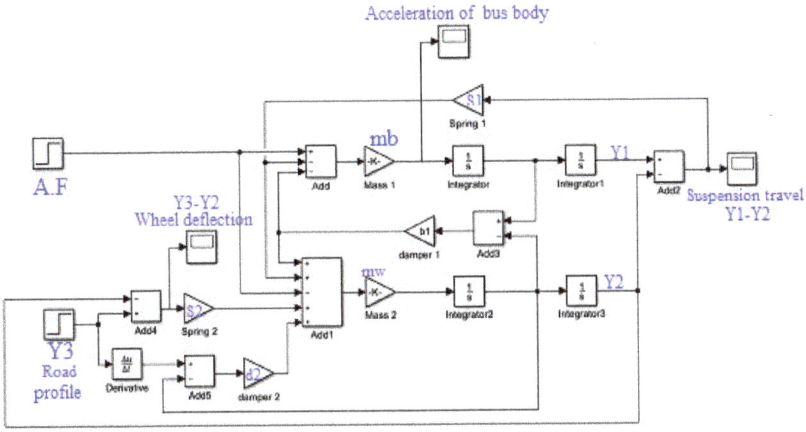

Figure 5. Simulink model of the bus's active suspension system.

Signals for Road Input

Two forms of road input signals will be employed to mimic various types of road conditions in this work. They are a bumpy road signal and a step input signal. These inputs are required to replicate the bus's suspension system, and they must precisely reflect the real-world road state when a vehicle travels on it. The Simulink model and the shape of each signal are shown in the four figures below (Figures 6–9).

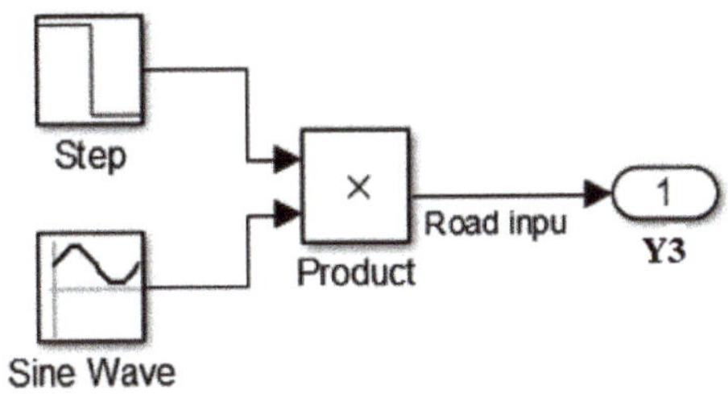

Figure 6. The bumpy road input's Simulink model.

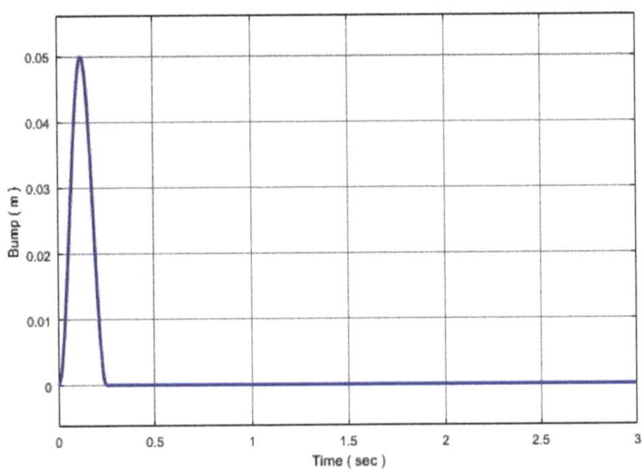

Figure 7. Bumpy road signal.

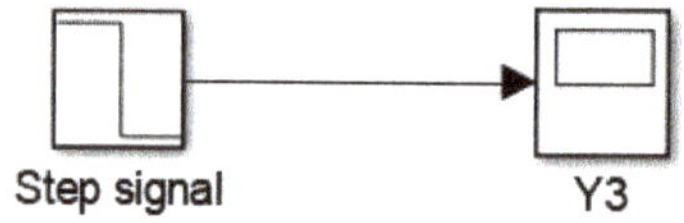

Figure 8. The step input's Simulink model.

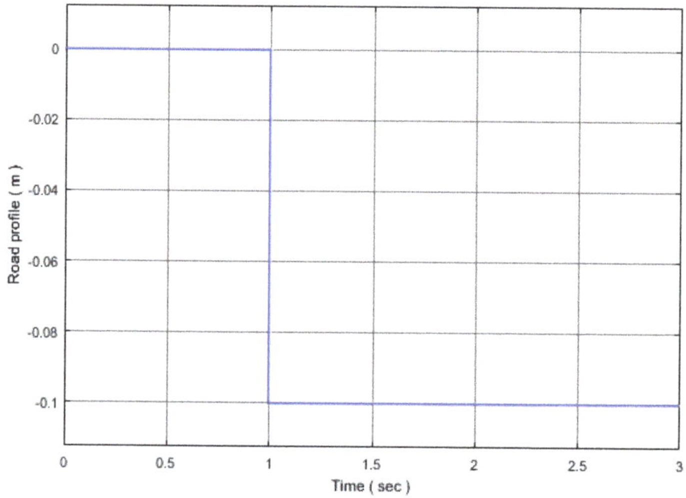

Figure 9. Unit step road signal.

3. Control Process

There are many studies related to the design and evaluation of the performance of suspension systems in vehicles, but the impact of the weight of in-wheel electric motors on the performance of the system has not been addressed in most of the published works. In this paper, one of the well-known control methods, linear quadratic control (LQR), will be used.

The linear quadratic regulator (LQR) is an optimum control that can be used for single-input single-output (SISO) or multi-input multi-output (MIMO) systems. It not only provides good stability but also has the potential to guarantee the system's stability margin [10]. It is worth noting that LQR can provide more optimal energy usage than PID and fuzzy controllers [11]. The linear quadratic regulator (LQR) is an ideal full-state feedback control law that regulates the control system by minimizing a quadratic cost function. Figure 10 illustrates the linear quadratic regulator controller schematic. As seen in Figure 11, we can see the block design of a full-state feedback controller utilizing LQR.

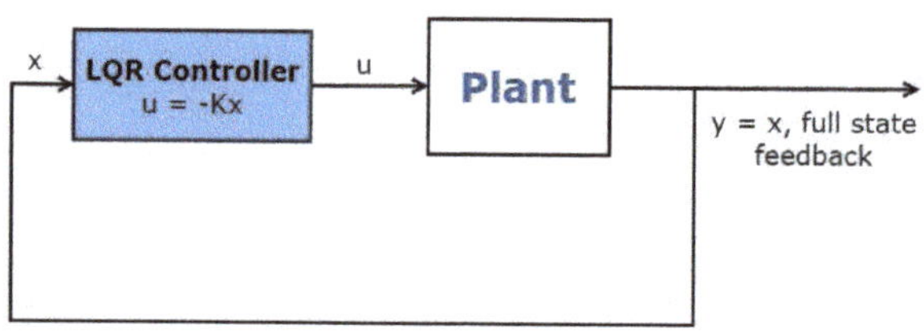

Figure 10. Schematic of the LQR controller.

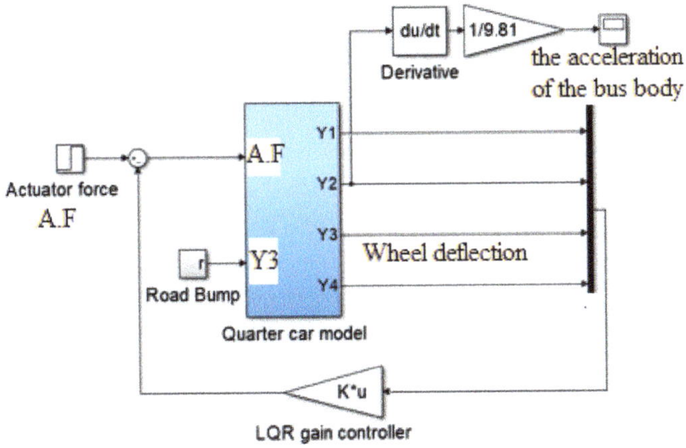

Figure 11. Active suspension system Simulink structure.

4. Simulations and Discussion

Through the suspension system shown in Figure 4, which represents a suspension system for a quarter of an electric bus, it consists of two main blocks, which are a quarter of the bus's mass, and the second block is the wheel mass, which is referred to by another term as the unsprung mass. The system also contains a set of springs and dampers as shown in the same figure. Using Newton's second law of motion, the two equations of motion were written for the system; based on the two equations, the Simulink model shown in Figure 5 was created. The inputs to the system are the actuator force and the road profile, and the main important outputs of the system are the acceleration of the bus's body and deflection of the wheel, as they refer to the passenger's comfort and road holding, respectively.

One of the well-known control methods was used, which has been explained in this paper, and it is the linear quadratic regulator, which was used on the basis of which control system was implemented, as shown in Figure 11. In this study, the behavior of the system was studied based on several variables, including the mass of the wheel (this was a standard wheel and an in-wheel motor) and the road profile (this was a step input and a bumpy road input).

Figures 12 and 13 denotes two parameters that affect the system's performance which are the wheel deflection and the acceleration of sprung mass with the step input, while Figures 14 and 15 show the same parameters but with the bumpy road input. The yellow lines in the four figures indicate the state of the system when using the standard wheels, while the blue lines indicate the system's performance when electric motors are used. We conclude that the use of in-wheel electric motors had a negative impact to some extent, and this is because these motors increase the weight of the wheels.

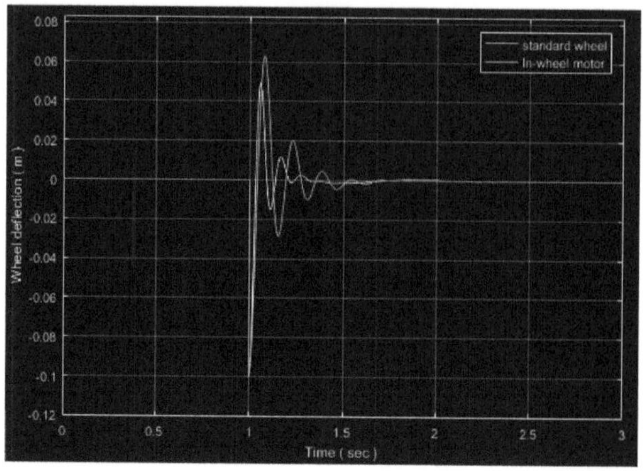

Figure 12. Wheel deflection with the step input.

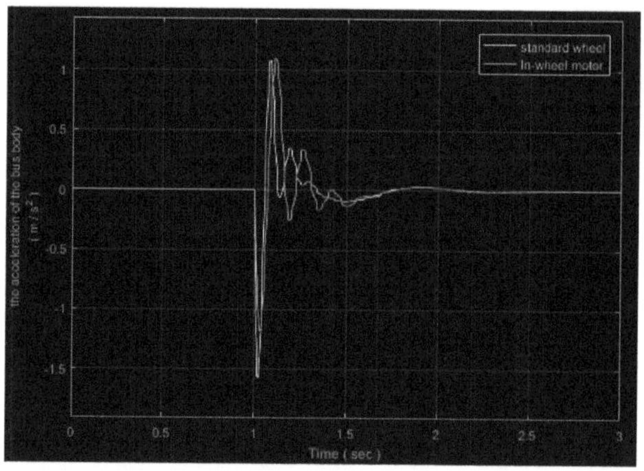

Figure 13. The acceleration of sprung mass with the step input.

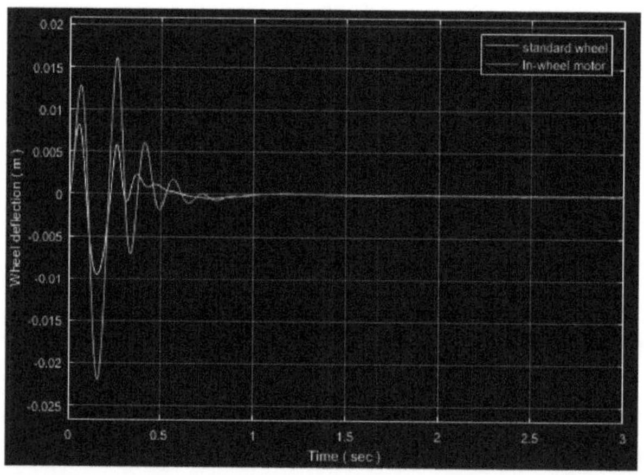

Figure 14. Wheel deflection with the bumpy road input.

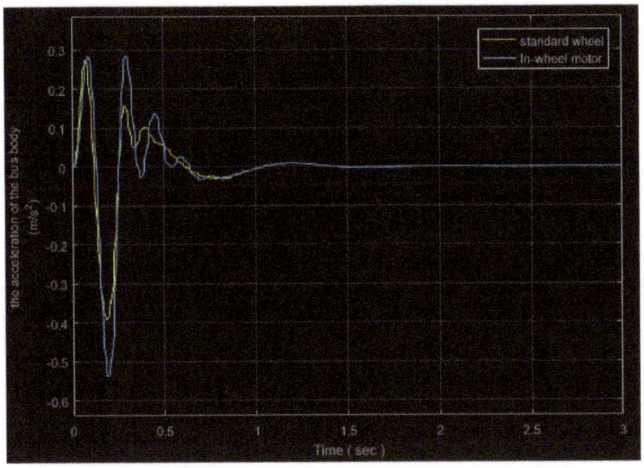

Figure 15. The acceleration of sprung mass with the bumpy road input.

5. Conclusions

The usage of an LQR controller to regulate the active suspension of an electric quarter bus is demonstrated in this study, as similar models are utilized in a large number of universities and institutions worldwide. It is critical that we are able to execute modeling in such a way that we may utilize a variety of different parameters and still obtain identical results. The main parameter that was taken into account was the weight of the bus's wheel, as it differs in this system. Once a standard wheel was used, a wheel with an electric motor was used again. This paper has shown that the electric motor increased the weight of the wheel, which negatively affected the performance of the system, so we need to use other control methods so that we may obtain better performance for the electric bus suspension system.

Author Contributions: Conceptualization, M.B. and A.A.A.; methodology M.B., A.A.A., M.M.K., A.A., M.M.R. and A.S.D.A.; formal analysis, A.A.A. and M.M.K.; investigation, M.B. and A.A.A. and A.A.; resources, all authors; data curation, all authors; writing—original draft preparation, M.B. and A.A.A.; writing—review and editing, all authors; visualization, all authors; supervision, all authors; project administration, A.A.A. All authors have read and agreed to the published version of the manuscript.

Funding: This research received no external funding.

Institutional Review Board Statement: Not applicable.

Informed Consent Statement: Not applicable.

Data Availability Statement: Not available.

Conflicts of Interest: The authors declare no conflict of interest.

References

1. Houbbadi, A.; Pelissier, S.; Trigui, R.; Redondo-Iglesias, E.; Bouton, T. Overview of Electric Buses deployment and its challenges related to the charging—The case study of TRANSDEV. In Proceedings of the 32nd Electric Vehicle Symposium (EVS32), Lyon, France, 19–22 May 2019.
2. Electric Buses: Where Are We? Available online: https://www.ies-synergy.com/en/electric-buses-where-are-we/ (accessed on 15 March 2021).
3. Global Electric Bus Adoption. Global Electric Bus Adoption to Triple by 2025. 2019. Available online: https://www.greentechmedia.com/articles/read/global-electric-bus-adoption-is-set-to-triple-by-2025 (accessed on 6 August 2021).
4. Kulkarni, A.; Ranjhal, S.A.; Kapoor, A. A quarter-car suspension model for dynamic evaluations of an in-wheel electric vehicle. *Proc. Inst. Mech. Eng. Part D J. Automob. Eng.* **2018**, *232*, 1139–1148. [CrossRef]

5. Salem, M.M.M.; Ayman, A.A. Fuzzy Control of a Quarter Car Suspension System. *World Acad. Sci. Eng. Technol.* **2009**, *53*, 1276–1281.
6. Ahmed, A.A.; Jomah, O.S.M. Modeling and Control of Car Active Suspension System Using a Neural Network-based Controller and Linear Quadratic Regulator Controller. In Proceedings of the 2020 IEEE 2nd International Conference on Electronics, Control, Optimization and Computer Science (ICECOCS), Kenitra, Morocco, 2–3 December 2020; pp. 1–6. [CrossRef]
7. Gordon, T.J.; Marsh, C.; Milsted, M.G. A Comparison of Adaptive LQG and Nonlinear Controllers for Vehicle Suspension Systems, Vehicle System. *Dynamics* **1991**, *20*, 321–340.
8. Ben Gaid, M.; Cela, A.; Kocik, R. Distributed control of a car suspension system. In Proceedings of the 5th EUROSIM Congress on Modeling and Simulation, Noisy-le-Grand, Franca, 6–10 September 2004.
9. Ahmed, A.A.; Emheisen, M. Analysis of Vehicle Handling Using a Simple Track Model of Automobile. In Proceedings of the 2019 19th International Conference on Sciences and Techniques of Automatic Control and Computer Engineering (STA), Sousse, Tunisia, 24–26 March 2019; pp. 130–133. [CrossRef]
10. Priyatmadi, A.P.; Sandiwan, H.; Wijaya, A.C. Application of spsa LQR tuning on quadrotor. In Proceedings of the 6th International Annual Engineering Seminar (IAES), Yogyakarta, Indonesia, 12–14 August 2016.
11. Maghfiroh, H.; Ataka, A.; Wahyunggoro, O.; Cahyadi, A.I. Optimal energy control of dc motor speed control: Comparative study. In Proceedings of the International Conference on Computer, Control, Informatics and Its Applications, Jakaarta, Indonesia, 19–20 November 2013.

engineering proceedings

Proceeding Paper

An Intelligent Optimization Algorithm for Scheduling the Required SIL Using Neural Network †

Naoual Batout *, Riad Bendib * and Youcef Zennir *

Department of Petrochemical and Process Engineering, University of Skikda, Skikda 21000, Algeria
* Correspondence: n.batout@univ-skikda.dz (N.B.); r.bendib@univ-skikda.dz (R.B.); y.zennir@univ-skikda.dz (Y.Z.)

† Presented at the 2nd International Conference on Computational Engineering and Intelligent Systems, Online, 18–20 November 2022.

Abstract: The purpose of safety analysis is to ensure that hazards and risks that could be a possible source of harm and damage are reduced well enough by dealing with all phases of the safety lifecycle and design of suitable safety barriers. It is known that any error or failure to perform the function of each proposed safety barrier can cause extreme damage to the environment, facilities and humans, and even loss of life. Therefore, it is necessary to ensure the effectiveness of the study or analysis. However, even with the major development in control system fields the problems of uncertainties, classification and optimization are still considered unsolved issues. In recent years several tools are developed based on artificial intelligence to deal with such difficulties. In this work, an approach based on Artificial Neural Networks (ANN) is developed to schedule the SIL values of the safety integrity functions (SIF) of an industrial-fired heater. The SIFs are first deduced from HAZOP study for the fired heater. The SIL risk of the consequences related to personnel health and safety, the economic SIL and environment SIL are considered as inputs of the multilayer network with a predefined hard limit activation function.

Keywords: optimization; ANN; hard limit; Safety; SIL; HAZOP; fired heater

Citation: Batout, N.; Bendib, R.; Zennir, Y. An Intelligent Optimization Algorithm for Scheduling the Required SIL Using Neural Network. *Eng. Proc.* **2023**, *29*, 5. https://doi.org/10.3390/engproc2023029005

Academic Editors: Abdelmadjid Recioui, Hamid Bentarzi and Fatma Zohra Dekhandji

Published: 11 January 2023

1. Introduction

Artificial Intelligence has a broad variety of application some of which we already know and encounter in our everyday life: spam filters recognizing malicious emails, search engine filters finding the "best results", vacuum cleaner robots or even no playable characters in video games [1,2].

The assumption that the human brain may be deemed quite comparable to computers in some ways offers the spontaneous basis for artificial intelligence (AI) [3,4].

The concept of AI was introduced following the creation of the notion of the Information Technology (IT) revolution, and is an attempt to replace human intelligence with machine intelligence [5]. According the Oxford dictionary, the word intelligence is derived from intellect, which is the faculty of knowing, reasoning and understanding. Intelligent behavior is, therefore, the ability to reason, plan and learn, which in turn requires access to knowledge.

AI requires a myriad of techniques, the most important of which is:

✓ artificial neural networks that rely on recognition system based on machine learning/deep learning to perform learning from observational data and discover their solutions [6].

2. Artificial Neural Networks

Artificial neural networks (ANNs) set out to emulate their biological equivalent. The simple model of neuron was proposed by MCCULLOCH and PITTS (1943), and HEBB

(1949) described a technique that became known as 'HEBBIAN learning'. ROSNBLATT (1961) developed a single layer of neurons called perceptron, which was used for optical pattern recognition [7].

WIDROW and SMITH (1964) purposed the first applications of this technology for control purposes. They developed an adaptive linear element (ADLINE) that was taught to stabilize and control an inverted pendulum. The back propagation training algorithm was investigated by WERBOS (1974) and further developed by RUMELHART (1986) and others, leading to the concept of the multi-layer perceptron (MLP) [8].

Artificial neural networks have the following potential advantages for intelligence control:

- They learn from experience rather than by programming;
- They have the ability to generalize from given training data to unseen data;
- They are fast and can be implemented in real-time;
- They fail gracefully rather than catastrophically [9,10].

2.1. The Formal Neuron

A formal neuron simply performs a weighted sum of those inputs, adds a threshold to that sum, and passes the result through a transfer function (activation function) to obtain its output like Figure 1 indicates [11].

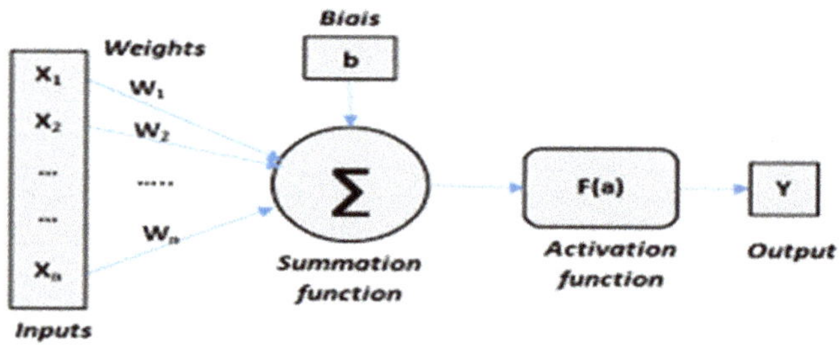

Figure 1. Formal neuron structure.

where:

$$Y = f(\sum_{j=1}^{n} wj\, xj - b) \quad (1)$$

2.2. Multi-Layer Networks

In this case, the networks generally have at least three layers, an input layer, one or more hidden layers and an output layer. Information flows from input to output through the hidden layer(s) as in Figure 2 [12].

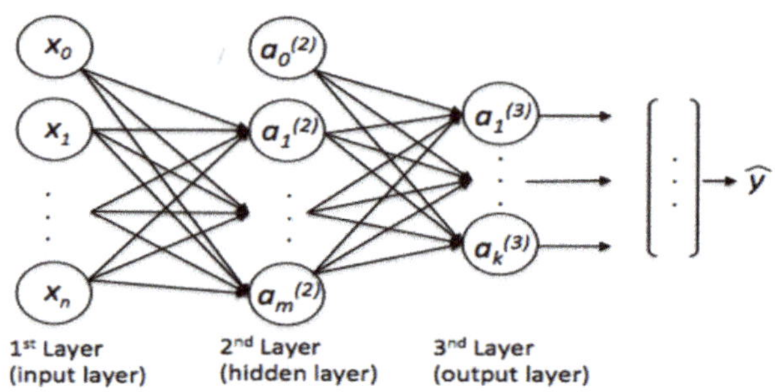

Figure 2. Multi-layer network structure.

2.3. Activation Function

Weighted input w and the bias b are summed up to create the transfer function net input, which is once more a scalar. This sum is the argument of the transfer function f.

f is a step function or a sigmoid function. Note that the neuron's scalar parameters w and b are both adjustable [1,12].

The fundamental concept behind neural networks is that these parameters can be changed to prompt the network to behave in an interesting or desired way. By adjusting the weight or bias parameters, we can instruct the network to perform a certain task. Alternatively, the network itself will adjust these parameters to achieve some desired end.

The shapes most used are presented in Figure 3.

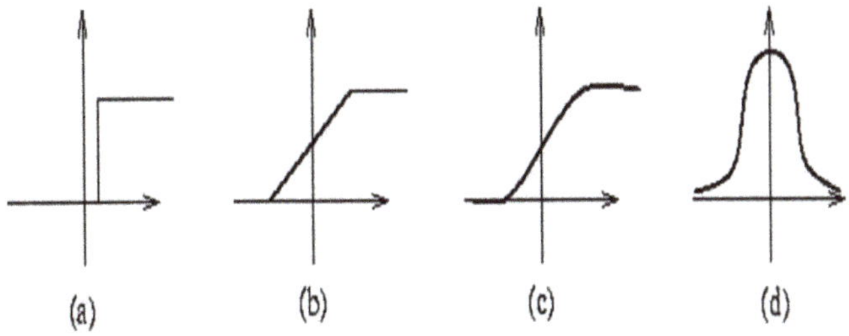

Figure 3. The function shape: (**a**) hard-limit function (Heaviside function) all or nothing (Neuron of Mac Culloch and Pitts (1945)); (**b**) linear function; (**c**) sigmoid function; (**d**) Gaussian function.

2.4. Neural Networks Learning

The learning consists of calculating the parameters in such a way that the network outputs are as close as possible to the "desired" outputs, which can be the code of the class to which the form that we want to classify belongs, the function value that we want to approximate or that of the process outputs that we want to model, or even the desired output of the process to be controlled [5,13].

Formal neural network learning techniques are optimization algorithms; they seek to minimize the gap between the network's actual responses and the desired responses by changing the parameters in successive steps (called "iterations"). The network output fits the data better and better as the training proceeds. However, the error made by the neural network at the end of learning is not zero [14,15].

Basically, there are two types of learning, unsupervised learning and supervised learning:

- Unsupervised learning: the algorithm must operate from unannotated examples. Indeed, in this case, machine learning is entirely independent. Data is then entered into the machine without being provided with examples of results. This mode of learning is also called competitive learning—letting the network self-organize by the local laws that govern the evolution of synaptic weights [16,17].
- Supervised learning: this is the most popular learning paradigm in machine learning and deep learning. As the name suggests, this consists of supervising the learning of the machine by showing it examples (data) of the task it must perform. The applications are numerous: Speech recognition, computer vision, regressions, classifications, etc. The vast majority of machine learning and deep learning problems use supervised learning [16,18].

3. Safety Integrity Level (SIL)

Safety instrumented systems (SIS) are technical systems that are widely used in the process industry. The mission of SIS is to detect the onset of hazardous events and to protect humans, material assets and the environment from their consequences. An SIS can perform several safety instrumented functions (SIF) and it is considered as an independent protection layer (IEC 61508 2010) [19].

A safety function is, thus, expressed in terms of the action to be taken and the required probability to satisfactorily perform this action [20].

As a quantitative target, this probability establishes the safety integrity [21]. The IEC 61508 defines four distinct safety integrity levels, SIL1, SIL2, SIL3 and SIL4, and the quantitative targets to which they are associated depend on whether the safety-related system is operating continuously or in low demand mode, for example, a shutdown system. The PFD or its inverse, the risk reduction factor, is the proper indicator of the first situation's safety function performance (RRF). Concerning the probability of a hazardous failure every hour is a function that runs constantly [22–24].

The four SIL definitions for low demand mode are shown in Table 1. As demonstrated, the more accessible the safety-related system is the higher SIL, and the more stringent becomes the implementation of safety function.

Table 1. Definitions of SILs for low demand mode.

SIL	Rang of Average PFD	Range of RRF
1	[10−2, 10−1]	[10, 100]
2	[10−3, 10−2]	[100, 1000]
3	[10−4, 10−3]	[1000, 10,000]
4	[10−5, 10−4]	[10,000, 100,000]

For determining the SIL, IEC standards have provided various methods that have been applied with differing degrees of success. These methods range from using pure quantitative risk assessments (QRAs) to more qualitative methods, as follows:

- Quantitative methods, such as fault tree analysis (FTA) and Markov graphs;
- Semi-qualitative methods, such as safety layer matrix, calibrated risk graph, and layers of protection analysis (LOPA);
- Qualitative methods, such as risk graph and hazardous event severity matrix.

The issue under this study is to classify the overall SIL for the deduced SIFs in HAZOP study (Table A1) [25].

The risk matrix used to identify SIL of different deduced SIFs takes into account the following: consequences related to environmental impact; consequences connected to production and equipment loss; consequences related to personnel's health and safety, is presented in Table 2 [20,26].

Table 2. Risk matrix.

Consequence Category			Demand Rate Category				
Health and Safety (S)	**Environmental (E)**	**Economic (L)**	**D0**	**D1**	**D2**	**D3**	**D4**
S0	E0	L0	-	-	-	-	-
S1	E1	L1	-	-	A1	A2	A2
S2	E2	L2	-	A1	A2	1	2
S3	E3	L3	-	A2	1	2	3
S4	E4	L4	-	1	2	3	4(x)
S5	E5	L5	-	2	3	4(x)	x

where: S0, … , S5 are categories of consequences on the health and safety of personnel (Table A2); E0, … , E5 are environmental consequences categories (Table A3); L0, … , L5 are economic consequence categories (Table A4).

4. Application

In this paper, methodology based on artificial neural networks is presented for the fired heater F201-101 of the crude distillation unit of ADRAR refinery Algeria represented in Figure 4. The unit is a part of ADRAR refinery, located in the south of Algeria.

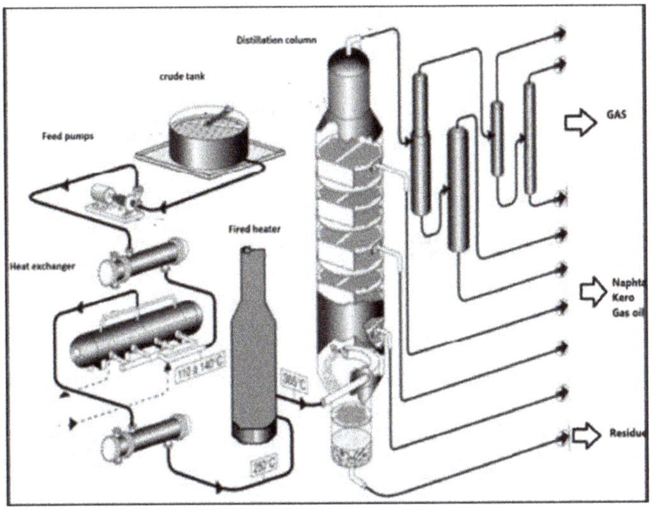

Figure 4. Process diagram for crude distillation unit.

Any fired heater should, in general, be controlled for the following parameters, as in Figure 5:

- The product flow on each heater pass (in our case the heater has two passes). Some heaters employ a combination of the flow in each pass and the skin temperature of the corresponding tubes to regulate the flow in the tube.
- The internal temperature of the tubes and the product. By using a cascade type controller, which regulates output temperature by pressure of fuel gas in the burners, it is possible to control the temperature of the product so that the set point is 365 c.
- The pressure: the pressure of the fuel gas in the burner's and the pilot's gas lines, as well as the pressure inside.

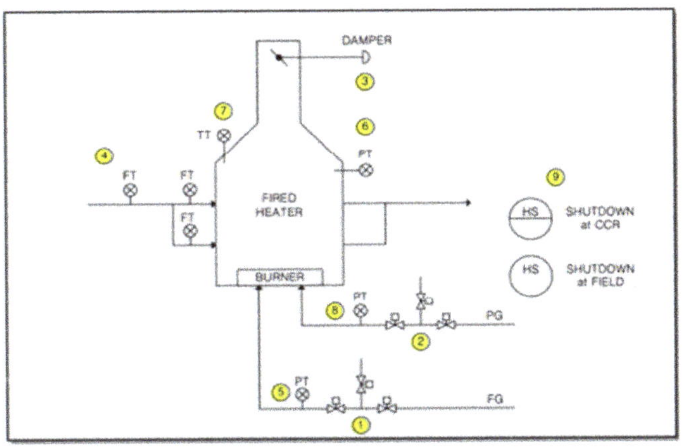

Figure 5. Process diagram for crude distillation unit.

The taken SIFs from [25], which were deduced based on the HAZOP study from [25] with their SILs, are included in the Table 3 (the possible scenarios in case SIF fails are summarized in Table A1.)

Table 3. SIL matrix values for each SIF.

	SIF 1/PS11203	
Economic Risk SIL	Environment Risk SIL	Personnel health SIL
2	(-) no safety requirement	(-) no safety requirement
	SIF 2/TAHH1	
Economic Risk SIL	Environment Risk SIL	Personnel health SIL
3	(-) no safety requirement	2
	SIF 3/FS11204, FS11205	
Economic Risk SIL	Environment Risk SIL	Personnel health SIL
3	(-) no safety requirement	2
	SIF 4/PAHH2	
Economic Risk SIL	Environment Risk SIL	Personnel health SIL
2	(-) no safety requirement	(-) no safety requirement
	SIF 5/TS11207	
Economic Risk SIL	Environment Risk SIL	Personnel health SIL
2	(-) no safety requirement	1
	SIF 6/PAHH3	
Economic Risk SIL	Environment Risk SIL	Personnel health SIL
3	(-) no safety requirement	2

The main objective for this study is to schedule the SILs values to the required SIL for the SIFs presented in Table 4, for this reason we have applied an optimization algorithm using a multi-layer artificial network (Figure 6).

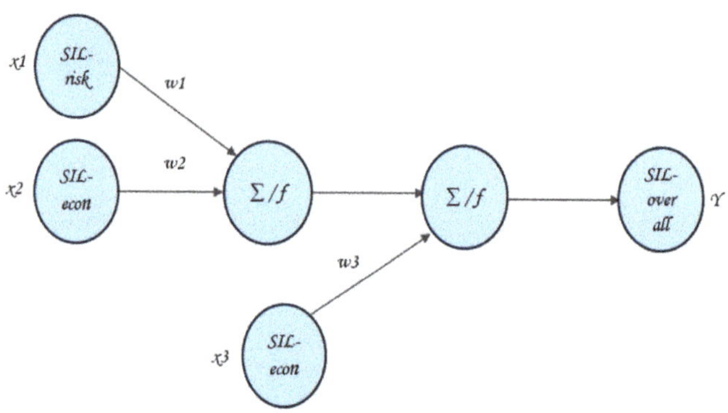

Figure 6. Network structure for the studied algorithm.

Inputs $x1$ and $x2$ to the neuron are multiplied by weights $w1$ and $w2$ and summed up together. The resulting n is the input to the activation function f. The activation function was originally chosen to be a relay function, but for mathematical convenience a hard-limit function is used; it is defined as

$$f = \begin{cases} x1 \; if \; wx > 0 \\ x2 \; if \; wx < 0 \end{cases} \quad (2)$$

The output of the first node becomes an input for the second node.

We used this function in our algorithm to create neurons that make classification decisions, and the typical network is shown in Figure 6.

The following table represents the network parameters.

Table 4. Network parameters.

Input Layer	Hidden Layer	Output Layer
$x1$: SIL-risk; $x2$: economic SIL;	$Y1$: output of first layer; $x3$: environment SIL	Y: SIL overall $F = \begin{cases} x1\ if\ wx > 0 \\ x2\ if\ wx < 0 \end{cases}$

5. Results and Discussion

The aim of this work is to create a cognition and decision system that classify the SILs values with a predefined activation function to define the overall SIL or the required SIL.

The work is conducted using MATLAB and results are presented in the below table (Table 5).

Table 5. SIFs deduced from HAZOP study.

SIF	$SIL_{overall}$
PS 11203	SIL2
TAHH 1	SIL3
FS11204 FS11205	SIL3
PAHH 2	SIL2
TS 11207	SIL2
PAHH 3	SIL3

As it is shown in the table, the safety integrity level of the heater's safety instrumented function are classified. The next step to ensure the safety of the fired heater is to compare the obtained results with the calculated SIL resulted from the calculation of the equivalent probability failure under demand of the corresponding safety integrity system. Depending on this comparison, recommendations for the safety system design are raised (i.e., keeping the existing component or proceeding to design configuration in case the calculated SIL is smaller than the required SIL) [25].

The parameters of the considered ANN are obtained during the learning step and they are suitable to be used in any complex system, as in the case of petrochemical plants [27].

Author Contributions: Conceptualization, N.B. and R.B.; methodology, N.B.; validation, N.B. and R.B.; formal analysis, N.B.; investigation, N.B.; resources, N.B.; data curation, R.B.; writing—original draft preparation, N.B.; writing—review and editing, N.B., R.B. and Y.Z.; visualization, N.B., R.B. and Y.Z.; supervision, R.B. and Y.Z.; All authors have read and agreed to the published version of the manuscript.

Funding: This research received no external funding.

Conflicts of Interest: The authors declare no conflict of interest.

Appendix A

Table A1. SIFs deduced from HAZOP study.

SIF	Definition	Type	Scenario
PS 11203	Low/low pressure in the fuel gas and pilot gas lines.	existing	Burner can extinguish low fuel gas pressure and possible flammable material accumulation
TAHH 1	Temperature High/High in tube	New	High/High temperature in the tube may lead to tube failure and explosion in case where the tube is damaged (pressure of hydrocarbons with fire). The existing protection is the low pressure vapour to extinguish the inside the heater.
FS11204 FS11205	Low/low flow of the crude in each pass	Existing	Low /low flow of the product in each pass will lead to increase the skin temperature of the corresponding heater tube which will lead to tube damage. Fire and explosion is expected.
PAHH 2	High/high alarm in the pressure of both fuel gas and pilot lines	New	Burner can extinguish at high/high fuel gas pressure as a result of gas blowing, and possible flammable material accumulation inside the heater. There is a possibility of explosion during heater restart-up.
TS 11207	High/high temperature in the heater box	Existing	Prolonged exposure to high temperature may cause tube failure which will lead to explosion and unit shutdown. High temperature of the crude may lead to perturbation of distillation column operation, and it may cause harm for the column internal in future.
PAHH 3	High/high pressure in the heater box	New	Increasing the pressure inside the heater box may lead to explosion. The existing physical protection is the explosion windows.

Table A2. Personnel safety and health categories.

Categories	Consequences
S0	No injury or health effect
S1	Slight injury or health effect
S2	Minor injury or health effect
S3	Major injury or health effect
S4	One to three fatalities
S5	Multiple fatalities

Table A3. Environment consequences categories.

Categories	Consequences
E0	No effect
E1	Slight effect
E2	Minor effect
E3	Local effect
E4	Major effect
E5	Massive effect

Table A4. Economic consequences categories.

Categories	Consequences
L0	No loss
L1	Slight loss < 10 K USD
L2	Minor loss 10–100 K USD
L3	Local loss 0.1–1 M USD
L4	Major loss 1–10 M USD
L5	Extensive loss > 10 M USD

References

1. You, M.; Li, S.; Li, D.; Xu, S. Applications of Artificial Intelligence for Coal Mine Gas Risk Assessment. *Saf. Sci.* **2021**, *143*, 105420. [CrossRef]
2. Rahmanifard, H.; Plaksina, T. Application of Artificial Intelligence Techniques in the Petroleum Industry: A Review. *Artif. Intell. Rev.* **2019**, *52*, 2295–2318. [CrossRef]
3. Wittwehr, C.; Blomstedt, P.; Gosling, J.P.; Peltola, T.; Raffael, B.; Richarz, A.-N.; Sienkiewicz, M.; Whaley, P.; Worth, A.; Whelan, M. Artificial Intelligence for Chemical Risk Assessment. *Comput. Toxicol.* **2020**, *13*, 100114. [CrossRef] [PubMed]
4. Huang, T.-J. Imitating the Brain with Neurocomputer a "New" Way towards Artificial General Intelligence. *Int. J. Autom. Comput.* **2017**, *14*, 520–531. [CrossRef]
5. Abiodun, O.I.; Jantan, A.; Omolara, A.E.; Dada, K.V.; Mohamed, N.A.; Arshad, H. State-of-the-Art in Artificial Neural Network Applications: A Survey. *Heliyon* **2018**, *4*, e00938. [CrossRef]
6. Cheng, J.; Li, Q.S.; Xiao, R. A New Artificial Neural Network-Based Response Surface Method for Structural Reliability Analysis. *Probab. Eng. Mech.* **2008**, *23*, 51–63. [CrossRef]
7. Nabian, M.A.; Meidani, H. Deep Learning for Accelerated Reliability Analysis of Infrastructure Networks. *Comput.-Aided Civ. Infrastruct. Eng.* **2018**, *33*, 443–458. [CrossRef]
8. Paltrinieri, N.; Comfort, L.; Reniers, G. Learning about Risk: Machine Learning for Risk Assessment. *Saf. Sci.* **2019**, *118*, 475–486. [CrossRef]
9. Adler, R.; Akram, M.N.; Bauer, P.; Feth, P.; Gerber, P.; Jedlitschka, A.; Jöckel, L.; Kläs, M.; Schneider, D. Hardening of Artificial Neural Networks for Use in Safety-Critical Applications—A Mapping Study. *arXiv* **2019**, arXiv:1909.03036.
10. Goh, Y.M.; Chua, D. Neural Network Analysis of Construction Safety Management Systems: A Case Study in Singapore. *Constr. Manag. Econ.* **2013**, *31*, 460–470. [CrossRef]
11. Anastasi, S.; Madonna, M.; Monica, L. Implications of Embedded Artificial Intelligence-Machine Learning on Safety of Machinery. *Procedia Comput. Sci.* **2021**, *180*, 338–343. [CrossRef]
12. Davoudi Kakhki, F.; Freeman, S.A.; Mosher, G.A. Use of Neural Networks to Identify Safety Prevention Priorities in Agro-Manufacturing Operations within Commercial Grain Elevators. *Appl. Sci.* **2019**, *9*, 4690. [CrossRef]
13. Albawi, S.; Mohammed, T.A.; Al-Zawi, S. Understanding of a Convolutional Neural Network. In Proceedings of the 2017 International Conference on Engineering and Technology (ICET), Antalya, Turkey, 21–23 August 2017; pp. 1–6.
14. Bengio, Y. Learning Deep Architectures for AI. *Found. Trends®Mach. Learn.* **2009**, *2*, 1–127. [CrossRef]
15. Kumar, M.; Jindal, M.K.; Sharma, R.K. Weka-Based Classification Techniques for Offline Handwritten Gurmukhi Character Recognition. In *Proceedings of the Second International Conference on Soft Computing for Problem Solving (SocProS 2012), 28–30 December 2012*; Babu, B.V., Nagar, A., Deep, K., Pant, M., Bansal, J.C., Ray, K., Gupta, U., Eds.; Springer: New Delhi, India, 2014; pp. 711–720.
16. Hertz, J.A. *Introduction To The Theory Of Neural Computation*; CRC Press: Boca Raton, FL, USA, 2019; ISBN 978-0-429-49966-1.
17. Fern´ndez, A.S.; Delgado-Mata, C.; Velazquez, R. Training a Single-Layer Perceptron for an Approximate Edge Detection on a Digital Image. In Proceedings of the 2011 International Conference on Technologies and Applications of Artificial Intelligence, Washington, DC, USA, 11–13 November 2011; pp. 189–193.
18. Carpenter, G.A.; Grossberg, S. A Self-Organizing Neural Network for Supervised Learning, Recognition, and Prediction. *IEEE Commun. Mag.* **1992**, *30*, 38–49. [CrossRef]
19. Fae, E.; Marcel, F. Standard Devices vs IEC 61508 SIL Safety Devices Integration in Process Industry. *Annu. Electr. Autom. Knowl. Shar. Event* **2018**, *15*, 199–207.
20. Bendib, R.; Zennir, Y.; Mechhoud, E.-A.; Bouziane, S. Risk Assessment for a Steam Generator (1050 G1) Skikda Refinery Algeria, Using HAZOP and RQA Methods. In Proceedings of the 2019 International Conference on Advanced Systems and Emergent Technologies (IC_ASET), Hammamet, Tunisia, 19–22 March 2019; pp. 262–267.
21. Nait-Said, R.; Zidani, F.; Ouzraoui, N. Fuzzy Risk Graph Model for Determining Safety Integrity Level. *Int. J. Qual. Stat. Reliab.* **2008**, *2008*, 263895. [CrossRef]
22. Cruz-Campa, H.J.; Cruz-Gómez, M.J. Determine SIS and SIL Using HAZOPS. *Process Saf. Prog.* **2010**, *29*, 22–31. [CrossRef]

23. Gabriel, A.; Ozansoy, C.; Shi, J. Developments in SIL Determination and Calculation. *Reliab. Eng. Syst. Saf.* **2018**, *177*, 148–161. [CrossRef]
24. Torres-Echeverria, A.C. On the Use of LOPA and Risk Graphs for SIL Determination. *J. Loss Prev. Process Ind.* **2016**, *41*, 333–343. [CrossRef]
25. Riad, B.; Hamid, B.; Hind, R.; Youcef, Z. Design Of an Integration Frame HAZOP-SIL for Safety Optimization of a Fired Heater. In Proceedings of the 2018 International Conference on Electrical Sciences and Technologies in Maghreb (CISTEM), Algiers, Algeria, 29–31 October 2018; pp. 1–6.
26. Bendib, R.; Mechhoud, E.; Bendjama, H.; Boulksibat, H. Risk Assessment of a Gas Plant (Unit 30 Skikda Refinery) Using Hazop &Bowtie Methods, Simulation of Dangerous Scenarios Using ALOHA Software. *Alger. J. Signals Syst.* **2020**, *5*, 25–32. [CrossRef]
27. Ahn, J.; Chang, D. Fuzzy-Based HAZOP Study for Process Industry. *J. Hazard. Mater.* **2016**, *317*, 303–311. [CrossRef] [PubMed]

engineering proceedings

Proceeding Paper

Design of Type 2 Fuzzy Logic Controller for FESTO Process Workstation †

Bendib Riad *, Mechhoud El-Arkam, Kassah Maroua, Kattouche Wiam and Zennir Youcef

Laboratory of Automatic, Department of Petrochemical, University 20 August 1955-Skikda, Skikda 21000, Algeria
* Correspondence: r.bendib@univ-skikda.dz
† Presented at the 2nd International Conference on Computational Engineering and Intelligent Systems, Online, 18–20 November 2022.

Abstract: Fuzzy logic is the most effective mathematical solution that has been presented in the last few years to deal with the problems in which imprecise and uncertain data exist. It can be defined as a generalization of classical binary logic, which admits only two logical states, true or false, by adding degrees of truth between the extreme values. Its basics were initiated by Lotfi Zadeh in the mid-1960s. However, in the last few years, a question was raised by different authors: yes, by using a fuzzy set or membership functions we can solve the problem of some kinds of uncertainties, but what is the situation in cases where the uncertainties exist in the membership function itself? To handle this situation, researchers introduce type 2 fuzzy logic. In our paper, we will introduce the design of a type 2 fuzzy logic controller to control the level in a FESTO process workstation. The mathematical model is first deduced, and after that the controller is designed. The simulation results show that the obtained controller gives very good transient characteristics for the system response work.

Keywords: fuzzy logic; level control; type 2 fuzzy logic; FESTO workstation

Citation: Riad, B.; El-Arkam, M.; Maroua, K.; Wiam, K.; Youcef, Z. Design of Type 2 Fuzzy Logic Controller for FESTO Process Workstation. *Eng. Proc.* **2023**, *29*, 6. https://doi.org/10.3390/engproc2023029006

Published: 11 January 2023

1. Introduction

One of the biggest problems generally facing chemical process industries (CPI) [1] is that they are very complex and accurate models are difficult, if not impossible, to obtain [2,3]. One reason for this is the nonlinear nature of most CPI problems. Another reason is the multivariable nature of control problems, since in almost all cases we need to control at least two parameters: the quantity and quality [4]. To overcome these problems, it will be helpful to apply techniques that use human judgment and experience rather than precise mathematical models, which in the major cases are deduced from the linearization of the system and simplification hypothesis. The fuzzy logic systems are capable of handling complex, nonlinear systems using simple solutions. Although they seem adequate for controlling chemical processes, since they provide solutions to incompletely defined and nonlinear processes, they are not widely used by researchers since obtaining an optimal set of fuzzy membership functions is not an easy task [5]. Another difficulty that faces the engineer when we use the fuzzy logic approach is this question: we believe that these techniques are suitable to handle uncertainties using the principle of membership degrees, but what is the situation in cases of the existence of uncertainties in the membership function itself? [6]. Different approaches are provided to deal with this situation, among them the use of what we call type 2 fuzzy logic [7]. In our work, we use a type 2 fuzzy logic controller to control the level in the FESTO MPS PA Compact Workstation [8,9]. A general description for this workstation is given in the following sections. To achieve the goal of our work, we first introduce a general description of fuzzy logic, followed by an introduction to type 2 fuzzy logic, process description and, finally, simulation and results.

2. Fuzzy Logic

2.1. Definition

The fuzzy logic tool was introduced in 1965 by Lotfi Zadeh [10,11], and is based on the concept of fuzzy sets. Fuzzy sets theory provides a mathematical tool for dealing with uncertainty. In general, fuzzy logic provides an inference structure that enables appropriate human reasoning capabilities. It is classified as an artificial intelligent tool. This is because fuzzy systems modeling, analysis, and control incorporate a certain amount of human knowledge into their components (fuzzy sets, fuzzy logic, and fuzzy rule base) [12,13]. In recent years, the number and variety of fuzzy logic applications have increased significantly. The applications range from consumer products such as cameras, camcorders, washing machines, and microwave ovens to industrial process control, medical instrumentation, decision-support systems, and portfolio selection.

2.2. Fuzzy Logic Control

Control applications are the kinds of problems for which fuzzy logic has the greatest success [14–16]. The basic structure of a fuzzy logic controller is shown in Figure 1.

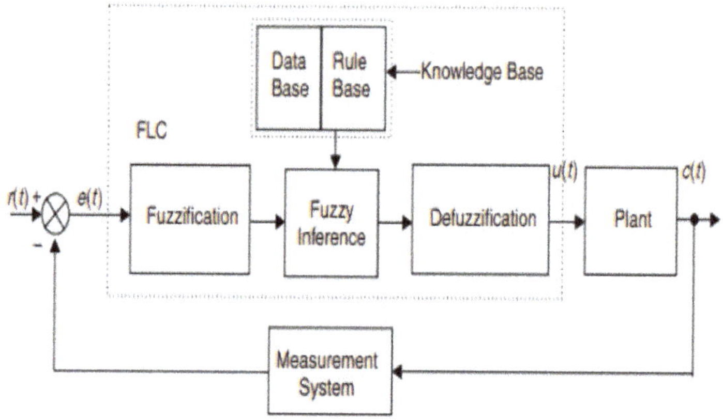

Figure 1. Fuzzy logic controller.

The measurements are processed using actual sensors (exact values), and after going through the fuzzification stage they transferred to linguistic variables, and in their fuzzier form, fuzzy inferences are computed, and the result is defuzzified; that is, it is transformed back into a specific number.

3. Type 2 Fuzzy Logic

The concept of a type 2 fuzzy set was introduced by Zadeh as an extension of the concept of an ordinary fuzzy set (henceforth called a type 1 fuzzy set) [17,18], in which the antecedent or consequent membership functions are type 2 fuzzy sets. Such sets are fuzzy sets whose membership grades themselves are type 1 fuzzy sets; they are very useful in circumstances where it is difficult to determine an exact membership function for a fuzzy set. They can be used to convey the uncertainties in membership functions of type 1 sets, due to the dependence of the membership functions on available linguistic and numerical information. Linguistic information (e.g., rules from experts), in general, does not give any information about the shapes of the membership functions. When membership functions are determined or tuned based on numerical data, the uncertainty in the numerical data, e.g., noise, translates into uncertainty in the membership functions. In all such cases, information about the linguistic/numerical uncertainty can be incorporated into the type 2 framework [15]. The structure of a T2FLS (type 2 fuzzy logic system), as presented in Figure 2, is quite similar to a type 1. The only difference is that the antecedent and/or consequent sets in a T2FLS are type 2. There are five principal parts in a T2FLS: fuzzifier, rule base, inference engine, type-reducer and defuzzifier.

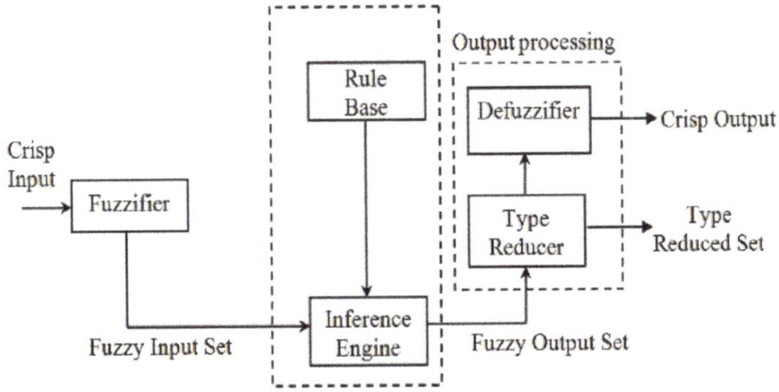

Figure 2. Type 2 fuzzy logic system.

4. Process Description (FESTO MPS PA Compact Workstation)

The FESTO MPS PA Compact Workstation shown in Figure 3 is a laboratory stand that has been used to design different kinds of PID controllers. As we can see in the photo, the stand is composed of a pump, pipes, two tanks, different types of sensors, actuators and a PLC controller S7-300 (Siemens Aktiengesellschaft, Nuremberg, Germany) to allow the testing of different types control systems. Using an OPC server algorithm, the PLC can communicate with the MATLAB software, in which the different control algorithms are designed. The system is equipped with a graphical tool implemented using WINCC software, to supervise the process and indicate the different measurements [19,20].

Figure 3. Type 2 Fuzzy logic system.

The FESTO Didactic station allows the realization of such a control system as: level, pressure, temperature and flow rate; in our case, the level control loop is used. The considered loop is shown in Figure 4. The process of the level control system is realized by the pump P101, that pumps liquid from tank B101 to tank B102 and the ultrasonic sensor located on the top of the tank B102. The measurement signal transmits the value of liquid in tank B102, and the correction is realized through an actuator signal which is the voltage applied to the pump. As is shown in Figure 4, the process admits a disturbance represented by manually opening/closing the valve V102 or the valve V110.

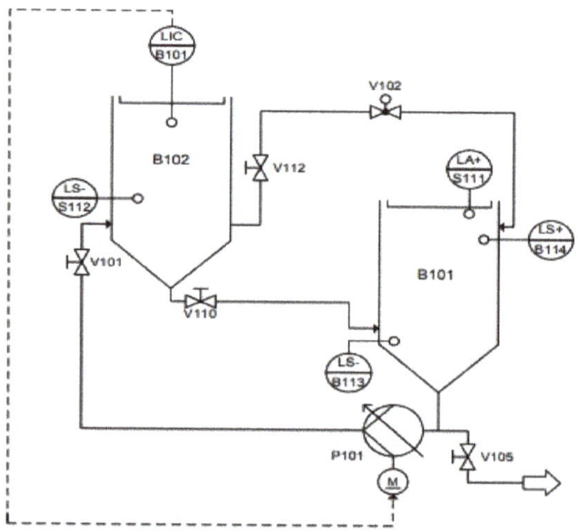

Figure 4. FESTO level control loop.

The best model that describes the relationship between the pump voltage and output flow rate was that identified, such that it provides better values of best fit for the training and validation tests. The transfer function given by is by Equation (1):

$$\frac{Y(z)}{U(z)} = \frac{0.013402\ z^{-1} - 0.013391\ z^{-2}}{1 - 1.970562\ z^{-1} + 0.970593\ z^{-2}} \tag{1}$$

5. Simulation and Results

In this section, the designs of type 1 and type 2 fuzzy logic controllers are presented.

5.1. Simple Type 1 Fuzzy Logic Controller

The Simulink Block diagram for the simple fuzzy logic controller applied for the FESTO workstation is shown in Figure 5, such that the inputs are the error and the variation of the error, and the output is the control action for the valve.

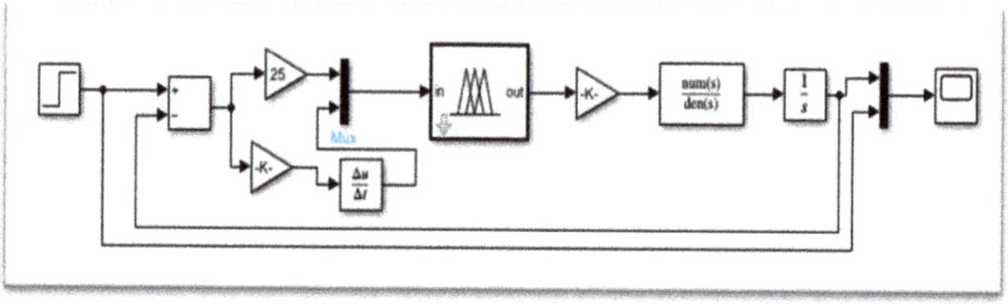

Figure 5. Simple fuzzy type 1 logic diagram applied for FESTO workstation.

The membership functions for the inputs and are shown in Figure 6.

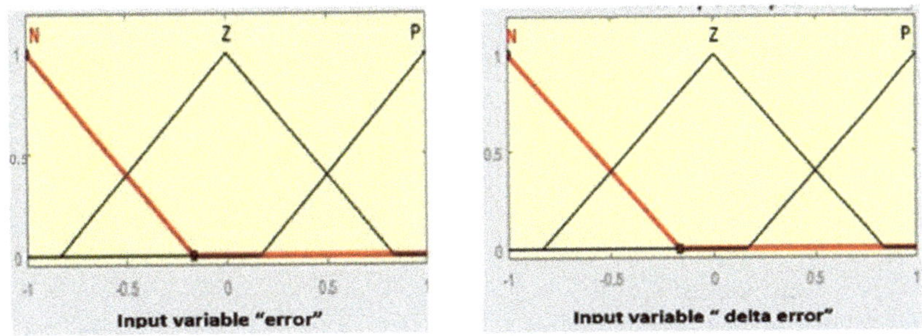

(**a**) (**b**)

Figure 6. The membership functions for the inputs. (**a**) Error; (**b**) Delta error.

Figure 7 shows the output membership functions for the control signal of the valve.

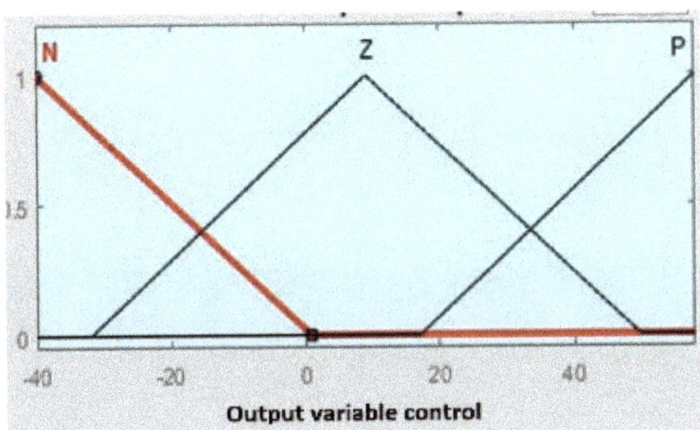

Figure 7. The membership function for the control action output.

The transient response for the system is shown in Figure 8:

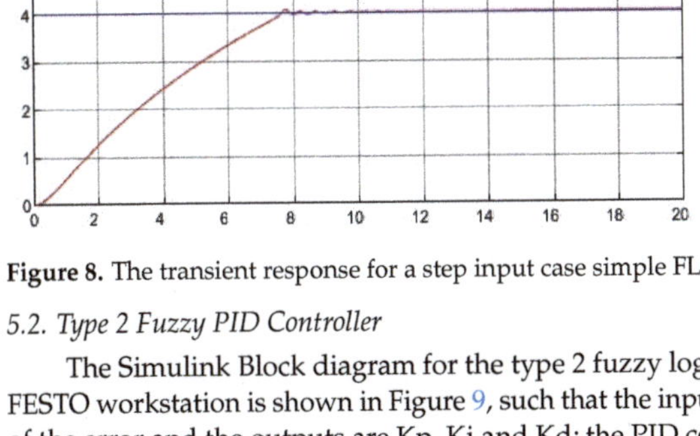

Figure 8. The transient response for a step input case simple FLC.

5.2. Type 2 Fuzzy PID Controller

The Simulink Block diagram for the type 2 fuzzy logic PID controller applied for the FESTO workstation is shown in Figure 9, such that the inputs are the error and the variation of the error and the outputs are Kp, Ki and Kd; the PID controller's parameters.

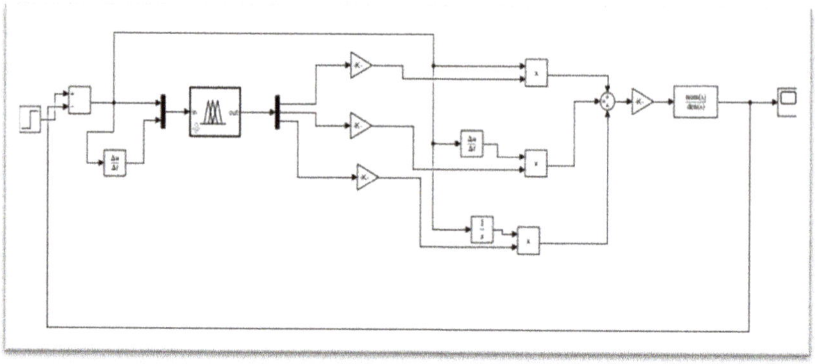

Figure 9. Simulink Block for Type 2 fuzzy logic PID controller.

The membership functions for the inputs are shown in Figure 10.

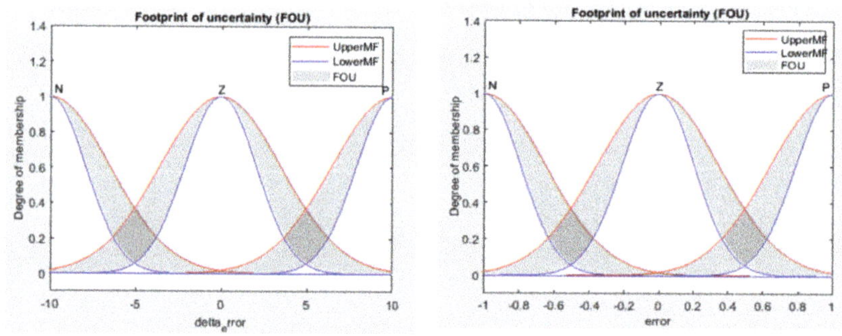

(**a**) (**b**)

Figure 10. The Type 2 membership functions for the inputs. (**a**) Error; (**b**) Delta error.

Figures 11–13 show the output membership functions of the PID parameters.

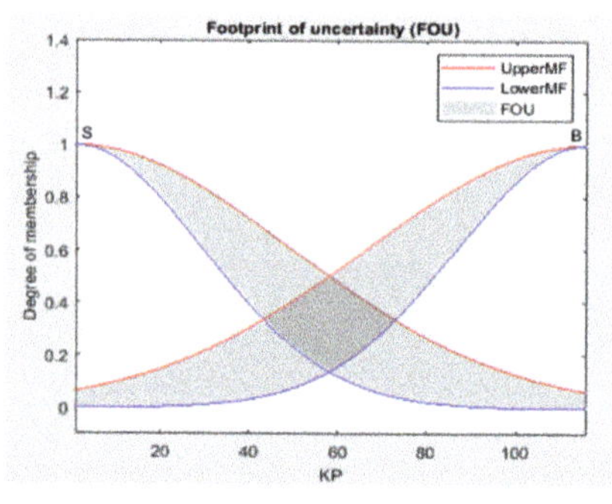

Figure 11. The membership function for KP.

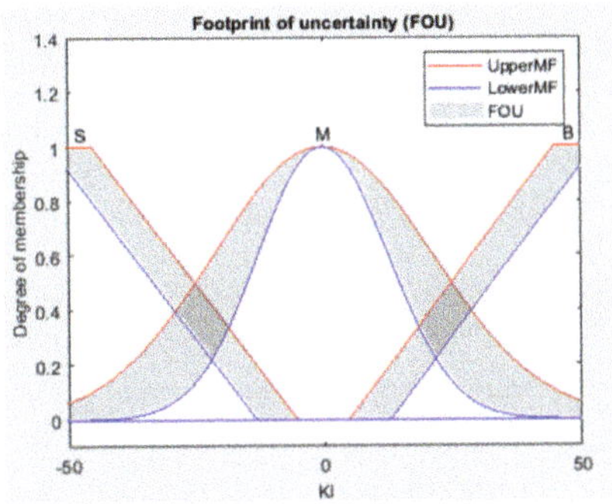

Figure 12. Ki membership function.

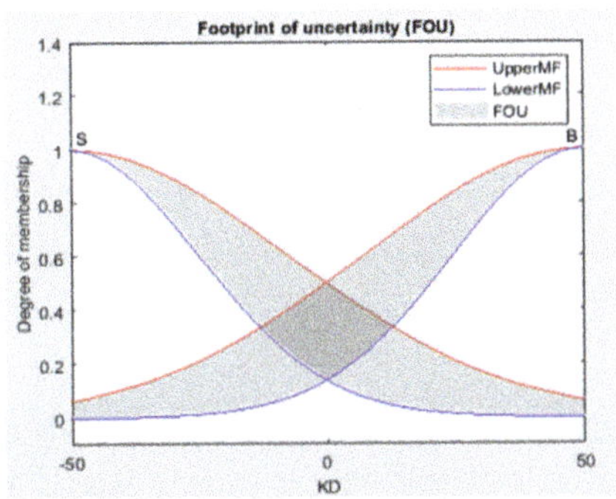

Figure 13. Kd membership function.

The transient response for the system is shown in Figure 14.

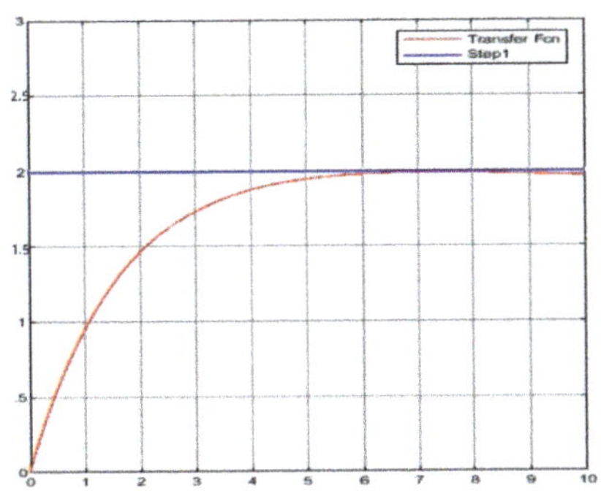

Figure 14. The transient response for a step input case Type 2 FLC.

The above figures (Figure 14) show that the system has a very fast response to the set point, with more precision and stability compared with the use of a simple fuzzy logic controller shown in Figure 8, the same remark for the use of simple PID control. The experimental and simulation results show that a better control in terms of robustness, speed and precision can be achieved by type 2 fuzzy logic controllers.

6. Conclusions

In this paper, we have presented the results of simulation with the two types of fuzzy control that we have studied in this work, using a simple fuzzy controller and also a fuzzy

PID controller for the FESTO Process Workstation. We have seen that the type 2 fuzzy controller shows very good performance when compared with the type 1, and from the results obtained we can say that the type 2 fuzzy controller is not just an extension, but an improvement of the type 1 fuzzy controller in the simulation framework.

Author Contributions: Conceptualization, B.R. and M.E.-A.; methodology, B.R.; software, K.M. and K.W.; validation, B.R.; formal analysis, B.R.; investigation, B.R.; resources, K.M. and K.W.; data curation, B.R.; writing—original draft preparation, B.R. and M.E.-A.; writing—review and editing, B.R.; supervision; Z.Y. All authors have read and agreed to the published version of the manuscript.

Funding: This research received no external funding.

Conflicts of Interest: The authors declare no conflict of interest.

References

1. Garvin, T.; Sinott, R. *Chemical Engineering Design, Principles, Practice and Economics of Plant and Process Design*; Elsevier: Amsterdam, The Netherlands, 2018.
2. Bendib, R.; Bentarzi, H. Contribution to Improve the Dynamic Control of a Binary Distillation Column. *Int. J. Syst. Assur. Eng. Manag.* **2017**, *8*, 663–675. [CrossRef]
3. Bendib, R. Optimization and Improvement of the Overall Performances of an Industrial Plant. Ph.D. Thesis, University of M'hamed Bougerra, Boumerdes, Algeria, 2017.
4. Skogestad, S.; Postlethwaite, I. *Multivariable Feedback Control: Analysis and Design*; John Wiley & Sons: Hoboken, NJ, USA, 2005; ISBN 978-0-470-01167-6.
5. Bendib, R.; Hammadi, Y.; Mazouzi, M.; Mechhoud, E. Particle Swarm Optimization for Tuning a Fuzzy Supervisory Controller Parameters (Takagi Seguno and Mamdani Engines). *Alger. J. Signals Syst.* **2020**, *5*, 92–97. [CrossRef]
6. Castillo, O.; Castro, J.R.; Melin, P. Type-2 Fuzzy Logic Systems. In *Interval Type-3 Fuzzy Systems: Theory and Design*; Castillo, O., Castro, J.R., Melin, P., Eds.; Studies in Fuzziness and Soft Computing; Springer International Publishing: Cham, Switzerland, 2022; pp. 5–11. ISBN 978-3-030-96515-0.
7. Karnik, N.N.; Mendel, J.M. Operations on Type-2 Fuzzy Sets. *Fuzzy Sets Syst.* **2001**, *122*, 327–348. [CrossRef]
8. Koszewnik, A.; Nartowicz, T.; Pawłuszewicz, E. Fractional Order Controller to Control Pump in FESTO MPS® PA Compact Workstation. In Proceedings of the 2016 17th International Carpathian Control Conference (ICCC), High Tatras, Slovakia, 29 May–1 June 2016; pp. 364–367.
9. Feijoo, J.D.; Chanchay, D.J.; Llanos, J.; Ortiz-Villalba, D. Advanced Controllers for Level and Temperature Process Applied to Virtual Festo MPS® PA Workstation. In Proceedings of the 2021 IEEE International Conference on Automation/XXIV Congress of the Chilean Association of Automatic Control (ICA-ACCA), Valparaíso, Chile, 22–26 March 2021; pp. 1–6.
10. Nguyen, H.T.; Walker, C.L.; Walker, E.A. *A First Course in Fuzzy Logic*, 4th ed.; Chapman and Hall/CRC: Boca Raton, FL, USA, 2018; ISBN 978-1-138-58508-9.
11. Timothy, J.R. *Fuzzy Logic with Engineering Applications*, 4th ed.; Wiley: Hoboken, NJ, USA, 2016. Available online: https://www.wiley.com/en-br/Fuzzy+Logic+with+Engineering+Applications%2C+4th+Edition-p-9781119235866 (accessed on 30 November 2022).
12. Kosko, B.; Tomos, M. *Fuzzy Thinking: The New Science of Fuzzy Logic*; Hyperion: New York, NY, USA, 1993.
13. Iron, I. *Fuzzy Logic-Controls, Concepts, Theories and Applications Logic with Engineering Applications*, 1st ed.; Dadios, E.P., Ed.; BoD–Books on Demand: Norderstedt, Germany, 2018.
14. King, E.R. *Computational Intelligence in Control Engineering*; CRC Press: Boca Raton, FL, USA, 1999.
15. Lee, C.C. Fuzzy Logic in Control Systems: Fuzzy Logic Controller. II. *IEEE Trans. Syst. Man Cybern.* **1990**, *20*, 419–435. [CrossRef]
16. Larkin, L.I. A Fuzzy Logic Controller for Aircraft Flight Control. In Proceedings of the 23rd IEEE Conference on Decision and Control, Las Vegas, NV, USA, 12–14 December 1984; pp. 894–897.
17. Zadeh, L.A. Fuzzy Sets. *Inf. Control* **1965**, *8*, 338–353. [CrossRef]
18. Zadeh, L.A. Outline of a New Approach to the Analysis of Complex Systems and Decision Processes. IEEE. 1973. Available online: https://ieeexplore.ieee.org/document/5408575 (accessed on 30 November 2022).
19. Sankata, B.P.; Umesh, C.P.; Kamalakanta, M. Implementation of Fuzzy-PID Controller to Liquid Level System Using LabVIEW. Available online: https://ieeexplore.ieee.org/document/6959045 (accessed on 30 November 2022).
20. Ahmad, S.; Ali, S.; Tabasha, R. The Design and Implementation of a Fuzzy Gain-Scheduled PID Controller for the Festo MPS PA Compact Workstation Liquid Level Control. *Eng. Sci. Technol. Int. J.* **2020**, *23*, 307–315. [CrossRef]

engineering proceedings

Proceeding Paper

Optimization of a Speed Controller of a WECS with Metaheuristic Algorithms †

Chetioui Lotfi [1], Zennir Youcef [2,*], Arabi Marwa [2], Horst Schulte [3], Mechhoud El-Arkam [2] and Bendib Riad [2]

1 LRPCSI Laboratory, Université 20 Aout 1955 Skikda, Skikda 21000, Algeria
2 Automatic Laboratory of Skikda, Université 20 Aout 1955 Skikda, Skikda 21000, Algeria
3 Control Engineering Group, Department of Engineering, University of Applied Sciences Berlin, 13353 Berlin, Germany
* Correspondence: y.zennir@univ-skikda.dz

† Presented at the 2nd International Conference on Computational Engineering and Intelligent Systems, Online, 18–20 November 2022.

Abstract: This paper describes the modeling and control of a high-power wind energy conversion system (WECS) using a variable speed doubly fed induction generator (DFIG) with the application of an MPPT method to obtain the maximum power from the system. We applied metaheuristic algorithms such as particle swarm optimization algorithms (PSO), Harris hawk optimization (HHO), and salp swarm algorithm (SSA) to optimize the speed sensor. The simulation results indicated that the MPPT method with the proposed optimized sensor could generate optimum rotor speed to achieve maximum power output. The simulation was developed using Matlab/Simulink.

Keywords: WECS-DFIG system; control design; metaheuristic algorithms (PSO, HHO and SSA algorithm)

Citation: Lotfi, C.; Youcef, Z.; Marwa, A.; Schulte, H.; El-Arkam, M.; Riad, B. Optimization of a Speed Controller of a WECS with Metaheuristic Algorithms. *Eng. Proc.* **2023**, *29*, 7. https://doi.org/10.3390/engproc2023029007

Academic Editors: Abdelmadjid Recioui, Hamid Bentarzi and Fatma Zohra Dekhandji

Published: 11 January 2023

1. Introduction

In recent decades, environmental pollution has been increasing, which is worrying, and conventional energy resources are being consumed rapidly, so the world has been turning to renewable energy resources (RES) [1,2]. RES can cover the entire energy needs of the world. There are many renewable energy resources such as solar, hydroelectric, biomass and wind energy, that have a positive impact [3]. Recently, there has been a serious shortage of power generated by conventional resources, so wind power, with its advantages of low cost and clean energy, is the best choice to meet society's energy needs. Multiple generators may be used with wind turbines, such as DFIG and permanent magnet synchronous generators (PMSG). DFIG is preferred in this area because it has significant advantages [4]. Variable speed wind turbine (VSWT)-based DFIG is superior to others, in addition to its high performance, due to the light weight, low cost, and small capacity of the converters [2–5]. To ensure high performance and achieve maximum performance, it is necessary to use an MPPT method [5]. Several types of MPPT methods have been applied to extract maximum power from WECS, such as maximum speed ratio (TSR), and disturbance and observation (D&O). In this study, we are interested in speed servo control, which consists of adjusting the electromagnetic torque of the generator in order to fix the mechanical speed to a reference speed that allows maximum power to be obtained from the turbine, based on the TSR technique extract.

2. Modeling of the WECS

The wind turbine absorbs the kinetic energy of the wind and converts it into torque, which turns the rotor blades. Three factors determine the relationship between the wind energy and the mechanical energy recovered by the rotor: the density of the air, the surface area swept by the rotor, and the wind speed. Air density and wind speed are climatological parameters that depend on site-dependent climatological parameters [6]. The power coefficient Cp characterizes the aerodynamic efficiency of the system (Figure 1). It depends on the characteristics of the turbine (blade dimensions, speed ratio and blade tip angle). We used an empirically approximated expression for a wind turbine using the DFIG-type generator defined as follows [7]:

$$C_p = \left[0.5176\left(\frac{116}{\lambda'}\right) - 0.4\beta - 5\right]exp\left(\frac{-21}{\lambda'}\right) + 0.0068\lambda \quad (1)$$

with:

$$\lambda' = \frac{1}{\lambda + 0.08\beta} - \frac{0.035}{\beta^3 + 1} \quad (2)$$

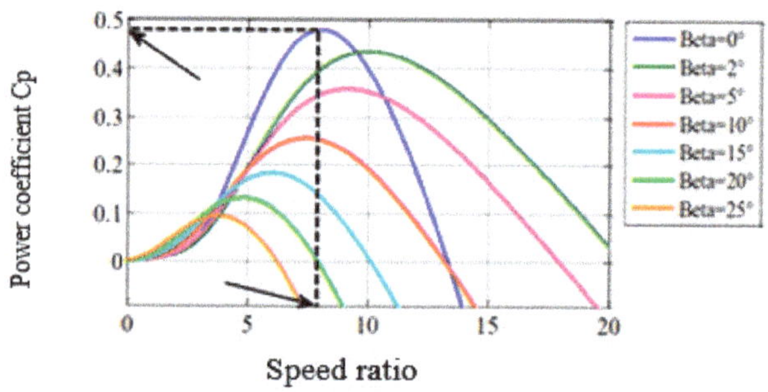

Figure 1. Power coefficient as a function of the speed ratio (λ) and stall angle (β).

In order to obtain the maximum power from the turbine (Figure 1), C_p and λ must have the optimal value for $\lambda = 8.1$ and $C_p = 0.48$. The mechanical power of the wind turbine and the aerodynamic torque is directly determined by:

$$P_t = 0.5C_p(\lambda, \beta)\rho\pi R^2 V^3 \text{ and } C_{aer} = \frac{P_t}{\Omega_t} \quad (3)$$

The gearbox is the connection between the turbine and the generator. It is used to match the highest speed of the generator to the slowest speed of the turbine. The figure below clarifies the structure of the wind energy conversion system (see Figure 2), and it is often modeled by the following two equations:

$$C_t = \frac{\Omega_{mec}}{G} \text{ and } C_g = \frac{\Omega_{aer}}{G} \quad (4)$$

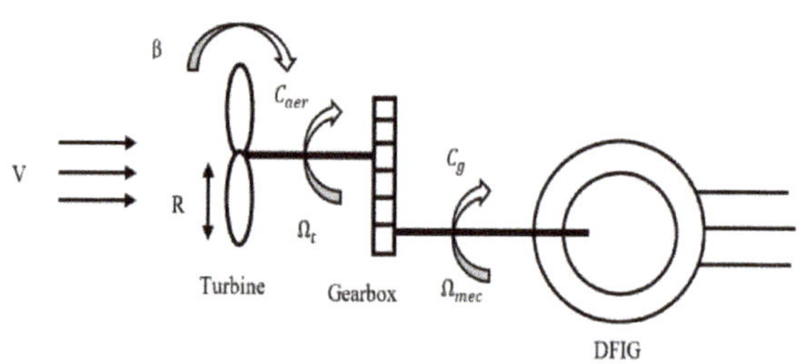

Figure 2. Structure of the wind energy conversion plant.

Modeling of the mechanical transmission can be represented as follows:

$$J\frac{d\Omega_g}{dt} = C_{mec} = C_g - C_{em} - C_f \quad (5)$$

The diagram block of the turbine of a horizontal axis variable speed wind turbine is shown in the following figure (Figure 3):

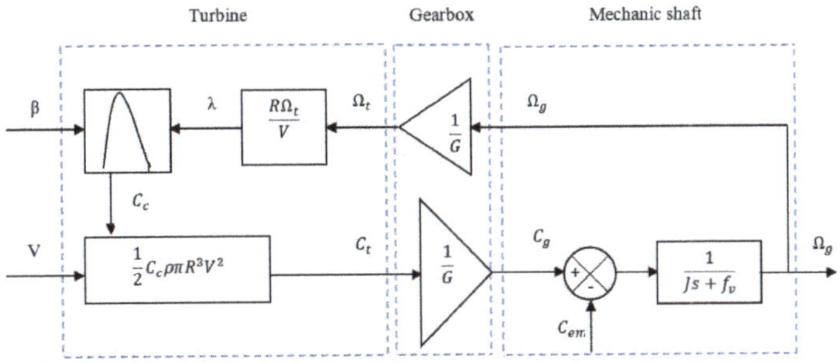

Figure 3. Block diagram of the turbine model.

The MPPT of the Proposed Wind Energy System

The power absorbed by the wind turbine can be maximized by adjusting the coefficient *Cp*. This coefficient depends on the speed of the generator. It possible to maximize this output using a variable speed wind turbine. It is, therefore, necessary to develop control strategies to maximize the generated electric power (thus, torque) by adjusting the speed of the turbine to its reference value regardless of the wind speed reference value considered as a disturbance variable [8]. The MPPT proposed in this work is the one with known airfoil properties with a control of the speed (the *Cp* is the one previously defined). This control structure consists of adjusting the torque appearing on the shaft of the Caer turbine in order to fix its speed to a reference value. In order to achieve this, the use of speed control (PI control) is absolutely necessary. The turbine speed reference is that which corresponds to the optimum value of the speed ratio *Cpmax* (=0, constant), which allows the maximum value of *Cp* to be obtained. Then we can write:

$$\Omega_{ref} = \frac{\lambda_{Cpmax} V}{R} \tag{6}$$

The installation shown in the figure below (Figure 4) represents the implementation of the MPPT with the wind turbine.

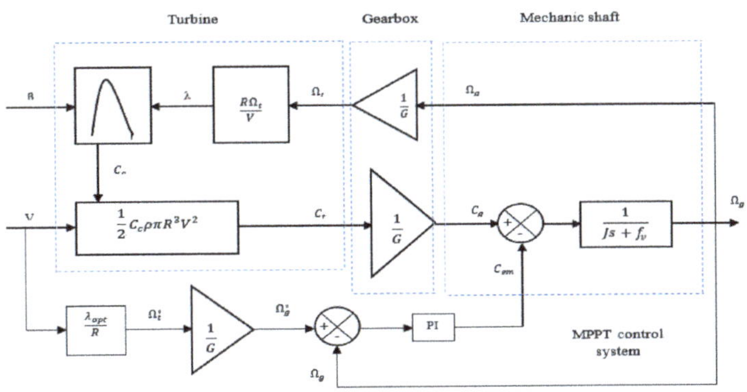

Figure 4. Principle of the MPPT proposed for the wind system.

3. DFIG Modeling

The DFIG is a machine that has excelled with its vector control. It is widely used in the wind turbine industry for variable speed wind turbines for various reasons such as the reduction of stress on the mechanical parts, the reduction of noise, and the possibility of controlling active and reactive power. The DFIG model in the *d-q* reference frame [7] is given by:

$$\begin{cases} V_{sd} = R_s i_{sd} + \frac{d}{dt}\varphi_{sd} - w_s\varphi_{sq} \\ V_{sq} = R_s i_{sq} + \frac{d}{dt}\varphi_{sq} - w_s\varphi_{sd} \\ V_{rd} = R_r i_{rd} + \frac{d}{dt}\varphi_{rd} - (w_s - w_r)\varphi_{rq} \\ V_{rq} = R_r i_{rq} + \frac{d}{dt}\varphi_{rq} - (w_s - w_r)\varphi_{rd} \\ C_{em} = \frac{3}{2}p\frac{L_m}{L_s}(\varphi_{qs} i_{dr} - \varphi_{ds} i_{qr}) \end{cases} \quad (7)$$

4. Proposed Optimization Algorithm

Metaheuristics are optimization procedures that make it possible to obtain an approximate value of the optimal solution in a reasonable amount of time. The goal is to solve a set of problems in different areas without having to change the basic principle of the algorithm of the method. In our work, we propose three metaheuristic algorithms—particle swarm optimization (PSO), Harris hawks optimization (HHO), and the salp swarm algorithm (SSA)—to see which is best in relation to our system.

4.1. Particle Swarm Optimization (PSO)

PSO is an optimization algorithm based on an evolutionary computational technique. The basic PSO was developed from swarming research such as fish flocking and bird flocking. After first being introduced in 1995, a modified PSO was introduced in 1998 to improve the performance of the original PSO. In PSO, social behavior is modeled by a mathematical equation that allows the particles to be guided during their displacement process [9–15]. The movement of a particle is influenced by three components: the inertial component, the cognitive component, and the social component. Figure 5 shows a flowchart of the method.

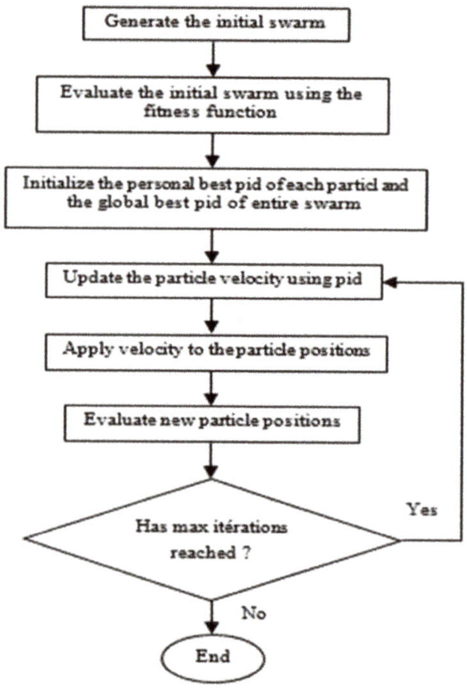

Figure 5. Flowchart of PSO.

4.2. Harris Hawks Optimization (HHO)

The cooperative hunting behavior of Harris hawks has inspired the HHO algorithm and has been mathematically modeled [11–14]. This algorithm is a population-based algorithm, which was inspired by nature and can be investigated under three main sections of exploration, the transition from exploration to exploitation, along with actual exploitation [14]. The flowchart of the HHO algorithm is shown in Figure 6. The exploration phase of the algorithm mimics the searching behavior of the hawks for prey. This is also the first stage of the algorithm. The mathematical model for this stage has a relation as given in Equation (8) [15]:

$$x(\tau+1) = \begin{cases} x_r(\tau) - r_1|x_r(\tau) - 2r_2x(\tau)| fork \geq 0.5 \\ (x_p(\tau) - x_m(\tau)) - r_3(L_B + r_4(u_B - L_B)) fork < 0.5 \end{cases} \quad (8)$$

where K, r_1, r_2, r_3 and r_4 are random numbers that have a range of [0, 1]. $x_p(\tau)$ stands for the prey's position, whereas the current and the next positions of the hawk are shown by $x(\tau)$ and $x(\tau+1)$, respectively. The random numbers of r_1, r_2, r_3 and r_4 are updated in each iteration. U_B and L_B are the search space's upper and lower bounds, respectively. $x_r(\tau)$ and x_m represent a randomly selected hawk in the available population and the average position of the population of hawks, respectively.

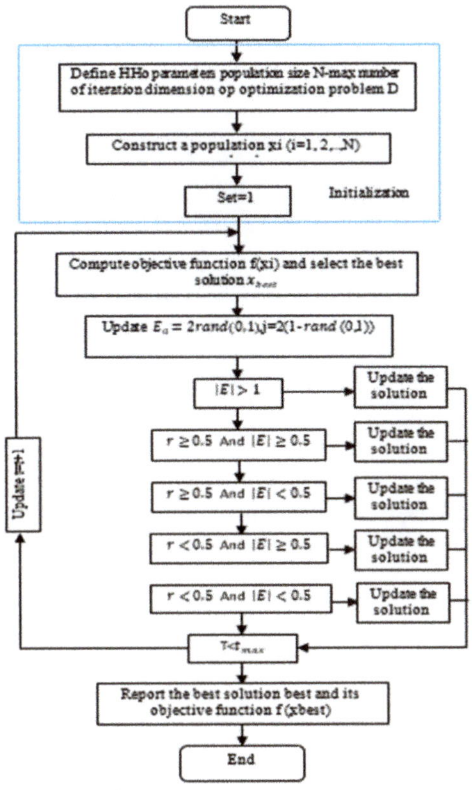

Figure 6. Flowchart of HHO.

4.3. Salp Swarm Algorithm (SSA)

SSA is one of the random population-based algorithms suggested by Mirjalili [12]. SSA simulates the swarming mechanism of salps when foraging in oceans. In heavy oceans, salps usually shape a swarm known as a salp chain. In the SSA algorithm, the leader is the salp at the front of the chain, and the remainder of salps are called followers. As with other swarm-based techniques, the position of the salps is defined in an s-dimensional search space, where s is the number of variables in a given problem. Therefore, the position of all salps is stored in a two-dimensional matrix called z. It is also assumed that there is a food source called P in the search space as the swarm's target. The mathematical model for SSA is given as follows: The leader salp can change position by using the next equation [12–14]:

$$z_n^1 = \begin{cases} P_n + r_1((U_n - l_n)r_2 + l_n) & r_3 \geq 0 \\ P_n - r_1((U_n - l_n)r_2 + l_n) & r_3 < 0 \end{cases} \tag{9}$$

where the meanings of all symbols are shown in Table 1.

$$r_1 = 2e^{-}\left(-\frac{4a}{A}\right)^2 \tag{10}$$

Table 1. The meanings of all symbols.

Symbols	Meaning
z_n^1	leader position in nth dimension
P_n	food source position in nth dimension
U_n	upper bound of nth dimension
l_n	lower bound of nth dimension
r_1, r_2, and r_3	random variables uniformly produced in the interval of [0, 1]
a	current iteration
A	maximum number of iterations
z_n^m	position of mith follower salp in nth dimension
e	time
v_0	initial speed

The coefficient r_1 is the essential parameter in SSA because it provides a balance between exploration and exploitation capabilities. To change the position of the followers, the next equations are utilized [16]:

$$z_n^m = \frac{1}{2}ce^2 + v_0 e \tag{11}$$

The flowchart of this algorithm is given as follows in Figure 7:

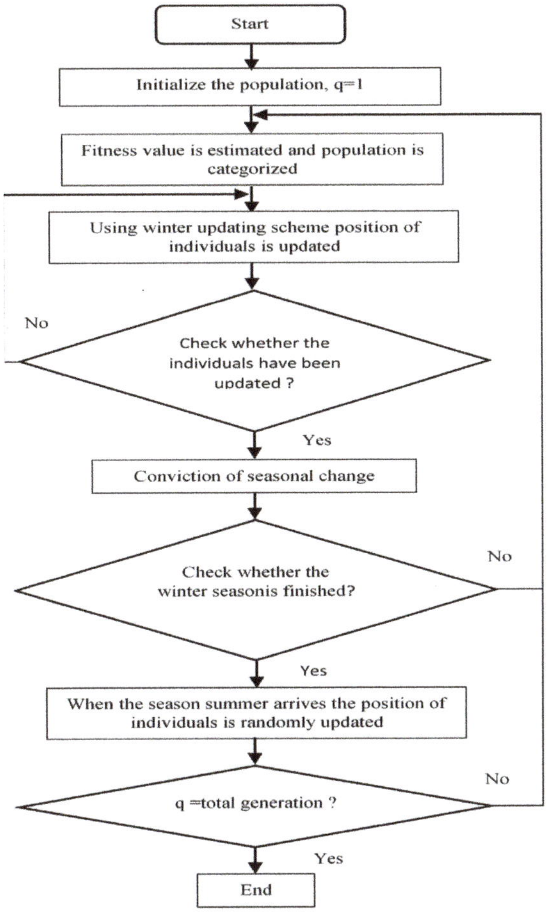

Figure 7. Flowchart of SSA.

5. Results and Discussion

In the simulation, the analysis was carried out in Matlab/Simulink. We present the results obtained for the dual-feed asynchronous generator (DFIG) with the application of the metaheuristic algorithm to optimize the parameters of the speed controller that provides the maximum power of the wind turbine; Figure 7 represents the block diagram used for the controller with the implementation of the applied optimization algorithms (Figure 8), which is widely used in the literature to evaluate the performance of PID controller design [13–15]. Rather than evaluating these error criteria as the objective alone, Zwe-Lee Gaing's study showed that objective combinations of these error criteria, which include eight factors, are a better way to form a common goal [15,16]. Many researchers use the method presented there. Four indices are commonly used to represent system performance: square integral error (ISE), absolute integral error (IAE), absolute temporal integral error (ITAE), and square temporal integral error (ITSE). They are defined as follows:

$$\begin{array}{c} IAE = \int |e(t)|dt \\ ITAE = \int t.|e(t)|dt \\ ISE = \int e(t)^2 dt \\ ITSE = \int t.\left|e(t)^2\right|dt \end{array} \quad (12)$$

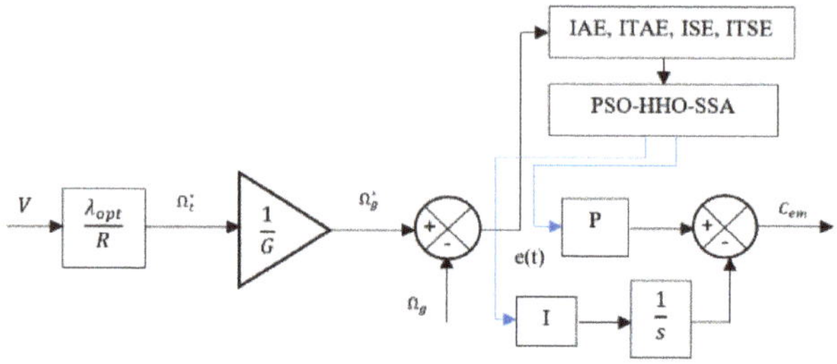

Figure 8. The block diagram of the implementation of the optimization algorithm for the speed turbine control.

The table below (Table 2) presents the different parameters obtained through the optimization, showing the best results for each algorithm after several tests.

Table 2. The different results obtained by optimization.

Algorithm	ISE	ITSE	IAE	ITAE	Error
PSO-Kp	19.3738	1.5885	23.2340	0.0001	0.4733
PSO-Ki	0.0001	28.5724	0.0001	17.9310	
HHO-Kp	50	50	37.0073	50	0.0023
HHO-Ki	34.9314	50	33.8005	30.0855	
SSA-Kp	50	50	50	34.2211	0.0003
SSA-Ki	29.9746	34.9321	3.1388	31.7244	

The colored values in the table are the best for minimum error for each algorithm used. The following figures show the evolution of various system parameters as a function of time for an average wind speed of 12 m/s and zero setting angle.

Table 2 presents the figures obtained by the most powerful algorithm, the SSA. We note that the SSA algorithm provided the optimal power coefficient Cp = 0.49 with a stall angle $\beta = 0$ (Figure 9) and a speed ratio $\lambda = 8.1$ as shown in Figure 10. Figure 11 shows the speed curve plotted with the reference speed, showing a minimum error of 3×10^{-3} and a stabilization time of 1.0274×10^5 s, where Figures 12 and 13 represent, respectively, the active and reactive power stabilizing at 100 s.

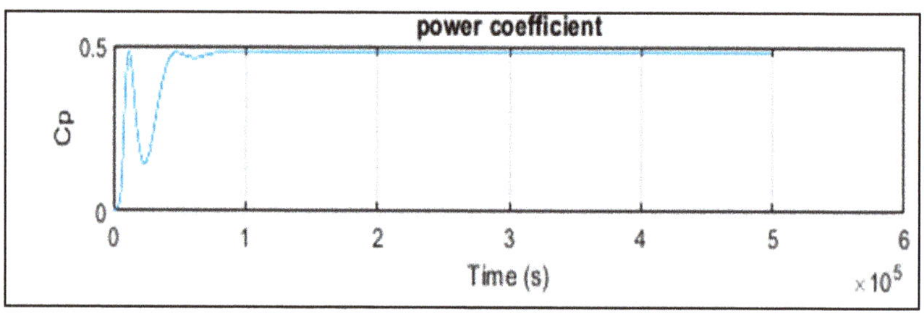

Figure 9. Power coefficient (with SSA).

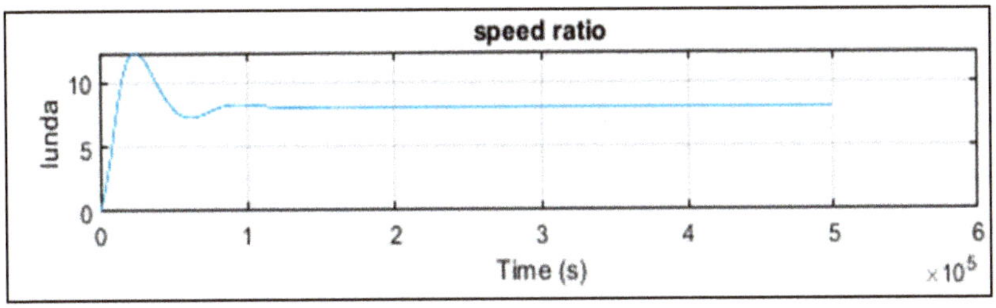

Figure 10. Speed ratio (with SSA).

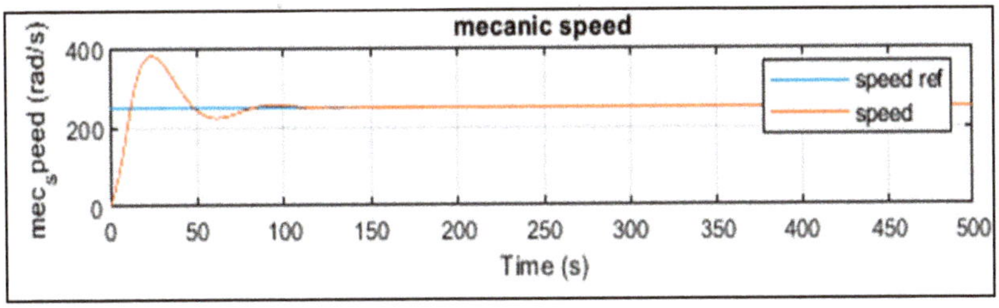

Figure 11. Mechanical speed (with SSA).

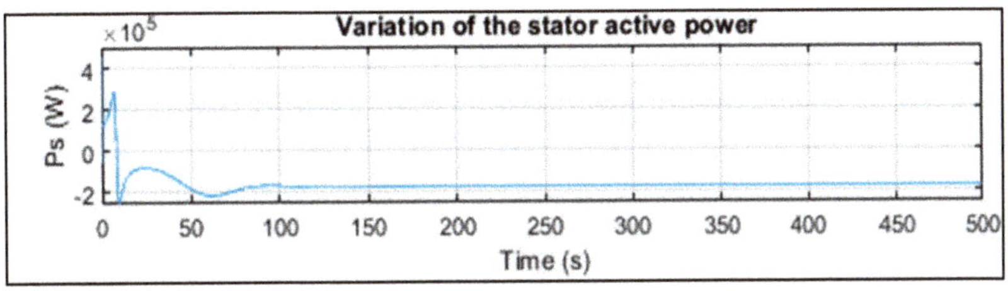

Figure 12. Variation of the stator active power (with SSA).

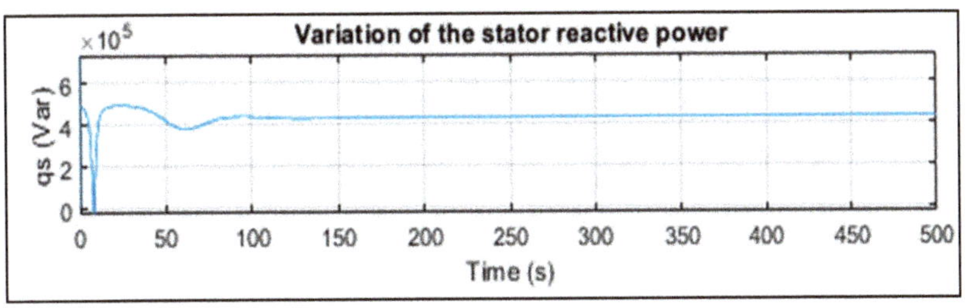

Figure 13. Variation of the stator reactive power (with SSA).

6. Conclusions

We proposed a full modeling of the VS-WECS in this paper. The system was based on a 1.5 MW DFIG, and the modeling and simulation of the wind turbine conversion chain were performed with a control architecture (MPPT) that maximized energy efficiency. Therefore, we played, on the one hand, on the efficiency of the conversion chain, and on the other hand, on the quality of the reactive power. To extract maximum power, we used a PI speed controller optimized with metaheuristic algorithms (PSO, HHO and SSA). The results obtained with the salp swarm approach (SSA) confirmed the effectiveness of our approach (stability and minimization of the wind turbine speed error). In future work, we will investigate the application of our approach to optimize a grid-connected wind farm.

Author Contributions: Conceptualization, C.L. and Z.Y.; methodology, C.L.; software, A.M.; validation, Z.Y. and H.S.; formal analysis, C.L., A.M. and Z.Y.; writing—original draft preparation, C.L., A.M., Z.Y., M.E.-A., B.R. and H.S.; writing—review, C.L., Z.Y. All authors have read and agreed to the published version of the manuscript.

Funding: This research received no external funding.

Data Availability Statement: Not applicable.

Conflicts of Interest: The authors declare no conflict of interest.

References

1. Sun, H.; Han, Y.; Zhang, L. Maximum Wind Power Tracking of Doubly Fed Wind Turbine System Based on Adaptive Gain Second-Order Sliding Mode. *J. Control Sci. Eng.* **2018**, *2018*, 1–11. [CrossRef]
2. Abdou, E.H.; Abdel-Raheem, Y.; Kamel, S.; Aly, M.M. Sensorless Wind Speed Control of 1.5 MW DFIG Wind Turbines for MPPT. In Proceedings of the Twentieth International Middle East Power Systems Conference (MEPCON), Cairo, Egypt, 18–20 December 2018; Cairo University: Giza, Egypt, 2018; pp. 700–704.
3. Kaloi, G.S.; Wang, J.; Baloch, M.H. Active and reactive power control of the doubly fed induction generator based on wind energy conversion system. *Energy Rep.* **2016**, *2*, 194–200. [CrossRef]
4. Yin, M.; Xu, Y.; Shen, C.; Liu, J.; Dong, Z.Y.; Zou, Y. Turbine Stability-Constrained Available Wind Power of Variable Speed Wind Turbines for Active Power Control. *IEEE Trans. Power Syst.* **2017**, *32*, 2487–2488. [CrossRef]
5. Falehi, A.D.; Rafiee, M. Maximum efficiency of wind energy using novel Dynamic Voltage Restorer for DFIG based Wind Turbine. *Energy Rep.* **2018**, *4*, 308–322. [CrossRef]
6. Mirecki, A. Etude Comparative de Chaînes de Conversion d'Energie Dédiées à une Eolienne de Petite Puissance. Ph.D. Thesis, Institut National Polytechnique de Toulouse, Toulouse, France, 2005; p. 252.
7. Djarir, Y. Commande Directe du Couple et des Puissances d'une MADA Associée à un Système Eolien par les Techniques de L'intelligence Artificielle. Ph.D. Thesis, Université Djilali LIABES Sidi-Bel-Abbès, Sidi Bel Abbès, Algeria, 2015; p. 285.
8. *Modeling, Identification and Control Methods in Renewable Energy Systems*; Springer Science and Business Media LLC: Marrakech, Morocco, 2019; pp. 1–374.
9. Sulaiman, D.R. Multi-objective Pareto front and particle swarm optimization algorithms for power dissipation reduction in microprocessors. *Int. J. Electr. Comput. Eng. (IJECE)* **2020**, *10*, 6549–6557. [CrossRef]
10. Rani, V.L.; Latha, M.M. Particle Swarm Optimization Algorithm for Leakage Power Reduction in VLSI Circuits. *Int. J. Electron. Telecommun.* **2016**, *62*, 179–186. [CrossRef]
11. Izci, D.; Ekinci, S.; Demiroren, A.; Hedley, J. HHO Algorithm based PID Controller Design for Aircraft Pitch Angle Control System. In Proceedings of the International Congress on Human-Computer Interaction, Optimization and Robotic Applications (HORA), Ankara, Turkey, 26–28 June 2020; p. 6.
12. Hegazy, A.E.; Makhlouf, M.A.; El-Tawel, G.S. Improved Salp Swarm Algorithm for Feature Selection. *J. King Saud Univ. Comput. Inf. Sci.* **2020**, *32*, 335–344. [CrossRef]
13. Kennedy, J.; Eberhart, R. Particle Swarm Optimization. In Proceedings of the IEEE International Joint Conference on Neural Networks, Perth, WA, Australia, 27 November–1 December 1995; pp. 1943–1948.
14. Heidari, A.A.; Mirjalili, S.; Faris, H.; Aljarah, I.; Mafarja, M.; Chen, H. Harris hawks optimization: Algorithm and applications. *Futur. Gener. Comput. Syst.* **2019**, *97*, 849–872. [CrossRef]
15. Gaing, Z.-L. A Particle Swarm Optimization Approach for Optimum Design of PID Controller in AVR System. *IEEE Trans. Energy Convers.* **2004**, *19*, 384–391. [CrossRef]
16. Kose, E. Optimal Control of AVR System with Tree Seed Algorithm-Based PID Controller. *IEEE Access* **2020**, *8*, 89457–89467. [CrossRef]

engineering proceedings

MDPI

Proceeding Paper

Risk Assessment in a Petrochemical Plant Using Socio-Technical Approaches (STAMP-STPA) †

Mechhoud El-Arkam *, Bendib Riad, Aribi Aya and Zennir Youcef

Laboratory of Automatic Skikda (LAS), Institut of Applied Sciences and Techniques, University 20 August 1955, Skikda 21000, Algeria

* Correspondence: el-arkam.mechhoud@univ-skikda.dz

† Presented at the 2nd International Conference on Computational Engineering and Intelligent Systems, Online, 18–20 November 2022.

Abstract: Major accidents continue to occur in chemical process industry, which may have serious consequences costing billion dollars and what is worse many human lives. That means the traditional Hazard analysis techniques are not becoming enough due to the increasing complexity of industrial plants. The main objective of this work is to present a new accident analysis technique based on system theory that has been developed lately changing the focus from reliability to system theory, in order to improve safety and for a better manage of risk. The considered is High Density Polyethylene plant which located in the Skikda industrial zone.

Keywords: STAMP; STPA; risk analysis; HDPE; petrochemical industries; safety; hazard

Citation: El-Arkam, M.; Riad, B.; Aya, A.; Youcef, Z. Risk Assessment in a Petrochemical Plant Using Socio-Technical Approaches (STAMP-STPA). *Eng. Proc.* **2023**, *29*, 8. https://doi.org/10.3390/engproc2023029008

Academic Editors: Abdelmadjid Recioui, Hamid Bentarzi and Fatma Zohra Dekhandji

Published: 11 January 2023

1. Introduction

There are many kinds of equipment in the petrochemical plants and petroleum refineries, usually presenting complex structures and several parameters. In such plants, it is important to consider different and critical types of risks, such as explosions, fire and toxic release which may cause serious damage either to human lives, equipment and plant or to the environment.

There are some traditional techniques developed during the last century employed to identify losses and hazards of chemical processes. The useful risk assessment methods are: HAZOP, FMEA, FTA, ETA … etc. [1], these methods are based on the component failures; they consider the risks as a chain of events and don't take in consideration the interaction between the components. Conventional methods have been used for a long time; they are not well-suited to handle modern systems with complex software, human-machine interactions, and decision-making procedures.

In order to conduct an automated risk assessment we are using another type of methodologies, as the socio-technical methods STAMP-STPA. In this paper we propose to implement the obtained results from STAMP-STPA, that is applied on the High Density Poly-Ethylene (HDPE) Reactor into a DCS to help the operator to take a good and safe decision.

2. Proposed Approach

The proposed approach is consist of four phases, the first one is the description of process, the second one is determining its position in the socio-technical system by STAMP method, the third phase is the application of System Theoretic Process Analysis (STPA) method and the fourth one is the implementation of the obtained results in the control system and display it in its HMI, to help operator in case of abnormal situation take a good and safe decision.

2.1. STAMP Systems Theoretic Accident Model and Processes

STAMP model is proposed by Leveson to model accident causation, it considers system theory as a useful way to analyze system accidents [2]. In STAMP safety is seen as a control problem, and it is managed by a control structure embedded in an adaptive socio-technical system, rather than in conventional methods, when accidents simply occur due to independent component failures, rather they occur when external disturbances, component failures, or dysfunctional interactions among system components are not adequately handled by the control system [2]. Safety management is defined as a continuous control task to impose the constraints necessary to limit system behavior to safe changes and adaptations. The main concepts in STAMP are: control loops, system model, constraints and levels of control. The cause of an accident, is viewed as the result of a lack of constraints imposed on the system design and on the operations, that is, by inadequate enforcement of constraints on behavior at each level of a socio-technical system, rather of a series of events [3].

2.2. STPA (System Theoritic Process Analysis)

STPA is a hazard analysis technique that embodies the STAMP accident causality model. It is based on control and system theory rather than the reliability theory underlying most existing hazard analysis techniques [4]. STPA does not generate a probability number related to the hazard. STPA is more powerful in terms of identifying more causal factors and hazardous scenarios, particularly those related to software, system design, and human behavior [5]. To apply STPA four steps should be followed:

1. Identify hazards and accidents: in this step we define the accidents and identify the hazards associated to those accidents.
2. Model the control structure: this step allow us to draw the feedback control system under study, from this model we can visualize actors involved, control actions and feedback information.
3. Identify unsafe control actions: Unsafe control actions UCAs are the wrong actions that lead to hazard. The UCAs are identified through four guide conditions taking advantage of control structure: (1) control action is not provided; (2) the UCA is provided; (3) control action is provided too late, too early or out of sequence; and (4) control action is stopped too soon or applied too long (applied only for continuous control). And Manuel et al added two UCAs (More and Less) because these two UCAs are adapted with petrochemical process, when we found the PID controllers [6].
4. Identify loss scenarios: this step is identified possible causes of UCAs causing hazards (losses). In this phase the information provided allows the engineers to change the design to eliminate or mitigate the causes of the hazards. This is the most important step of the analysis but also de most difficult one [5].

3. Case Study

In this section we study the HDPE reactor located in CP2K unit in polymed complex sonatrach-Skikda. The study consists of process description, the application of STPA method on this industrial process, in order to improve their safety and implement the obtained results in the control system in order to automated safety process [1].

3.1. Process Description

The HDPE reactor (Figure 1) is piping with length of 304 m and internal diameter of 560 mm in the form of loop, composed of four vertical sections, linked by horizontal sections. The vertical sections have Jacket insulated for heating in the phase of preparation and refrigeration in the phase of reaction. Those, made of carbon steel with external diameter 760 mm, are designed to the pressure and the temperature of 15 kg/cm^2g and 142 °C respectively. The reactor can be decomposed into the four following parts:

- The reactor enclosure
- The reactor pump
- Refrigeration system

- Decantation legs.

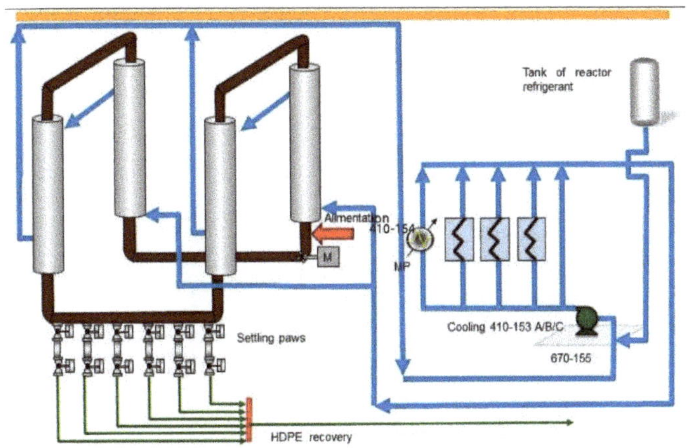

Figure 1. Polymerization process.

The reactor feed streams (ethylene, isobutane, hydrogen and 1-hexene, in the case of the production of copolymers) require a high degree of purity, for this; they are in advance treated to remove any catalyst poison (basically acetylene, oxygen, and water) until not harmful residual contents [7]. This is accomplished in suitable catalytic caterers, in the case of ethylene, degassing columns, isobutane and hexene-1, and specific dryers for all currents. The reactor is fed with the raw materials processed at the treatment area. Recycled isobutane, hydrogen, hexene-1 and ethylene arrive at the reactor through the main supply line to the reactor. Hexane and recycled isobutane are mixed in the static mixer isobutane/hexene. Hydrogen is mixed with the ethylene and it is added to the stream of recycled isobutane/hexene at the mixer output. The feed to reactor at different flows is adjusted based on certain variables. The isobutane-ethylene-polyethylene mixture flows into the reactor through the reactor pump [8].

3.2. Application of STAMP-STPA

Figure 2 shows the hierarchical control structure of polymerization plant, when we can identify the different unsafe control actions, the relation between each services and components and the nature of each information and order.

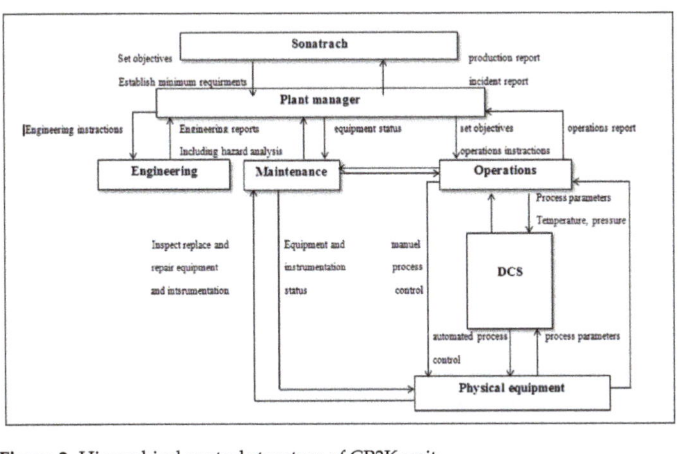

Figure 2. Hierarchical control structure of CP2K unit.

3.2.1. Hazards Identification

In our case study, we consider the variation of the temperature and pressure parameters, caused by the variation in the flow of the entering products (Ethylene, Hexene, and Isobutane), in addition the actions of cooling water valves and settling paws discharge valves. Table 1 shows the hazards identification and its constraints.

Table 1. Hazards identification.

Accident	Hazard	Safety Constraints
Explosion	H1: too high temperature	Temperature must never violate a maximum value
	H2: too high pressure	Pressure must never violate a maximum value

3.2.2. Draw the Control Structure

In this step we draw the control structure of the HDPE reactor in order to identify all the unsafe control actions that can lead to the two hazards identified in the precedent section, which are caused the explosion (Figure 3).

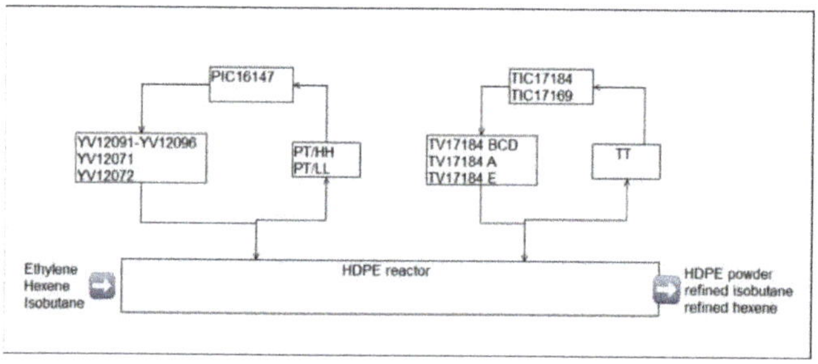

Figure 3. Control structure of HDPE reactor.

3.2.3. Identify Losse Scenarios

To determine the size of STPA table we use the following equasion:

$$STS = \prod_{i=0}^{N} number\ of\ UCAs \prod_{j=0}^{M} Number\ of\ stats \quad (1)$$

We choose (MORE (+), LESS (-), PROVIDED (1) and NOT PROVIDED (0)) as UCAs on each state. In this case, if we consider all the UCAs possible related to all states, we get:

$$STS = 4 \times 4 \times 4 \times 4 \times 4 = 4^5 = 1024 \quad (2)$$

So we get a big siize of the STPA table (Table 2) in term of rows number, in order to reduce the size of this table we consider only these UCAs: MORE, LESS, and NOT PROVIDEDconcerning the entering products (Ethylene and Hexane) and the cooling valve (for that it work continuously), PROVIDED (1) and NOT PROVIDED (0) concerning the flow of Isobutane and the opening of the discharging valves, in this case we get:

$$STS = 3 \times 3 \times 3 \times 2 \times 2 = 108 \quad (3)$$

From the STPA table we can see that the UCAs which lead to hazards are: more (+) concerning the entering flow of Ethylene and Hexene, and not provided (0) concerning the opening of the pressure valves (settling paws and safety valves). STPA analysis results not only in the detection of hazardous situations but also offer the technical solutions from the same table. In this case, it can be easily seen that all of those scenarios (colored green) corresponds to no hazards and the critical scenarios (colored red).

Table 2. STPA table.

ID	$F_{ethylene}$	F_{hexene}	Cooling Valve	$F_{isobutane}$	Settling Paw Valve	Hazard
1	+	+	+	0	0	H2
2	+	+	-	0	1	H1
3	+	+	0	1	0	H1, H2
4	+	-	+	1	1	No
5	+	-	-	0	0	No
6	+	-	0	0	1	H1
7	+	0	+	1	0	H2
8	+	0	-	1	1	H1
9	+	0	0	0	0	No
⋮	⋮	⋮	⋮	⋮	⋮	⋮
105	0	-	0	0	0	No
106	0	0	+	0	1	No
107	0	0	-	1	0	No
108	0	0	0	1	1	No

3.2.4. Recommendations

The recommendations have been offered to avoid the different hazardous situations are:

- Close the reactive flows when the cooling valves are not open.
- Open the pressure valves when the pressure in the reactor violate a maximum value because of the over flow of the reactives.
- Verify the timing settings concerning the opening of the discharging valves of the settling paws for that the high pressure could be easily discharged.
- Display textual messages on the HMI of DCS control room for that we ensure that the safety actions needed to be applied on the process can be easily done by any available operator in the control room in order to improve safety measures in the plant.

3.3. *Implementation Results in Control System*

In this stage we implement the obtained results from the STPA method into the control system to automate process safety, firstly we programme all cases in Tristation software (Triconnex PLC) (Figure 4), then we buld the HDPE reactor HMI (Figure 5) into Intouch software to monitor our process and display the alarms and abnormal situations, all these operations are mad in order to help the operators to take good and safe decision.

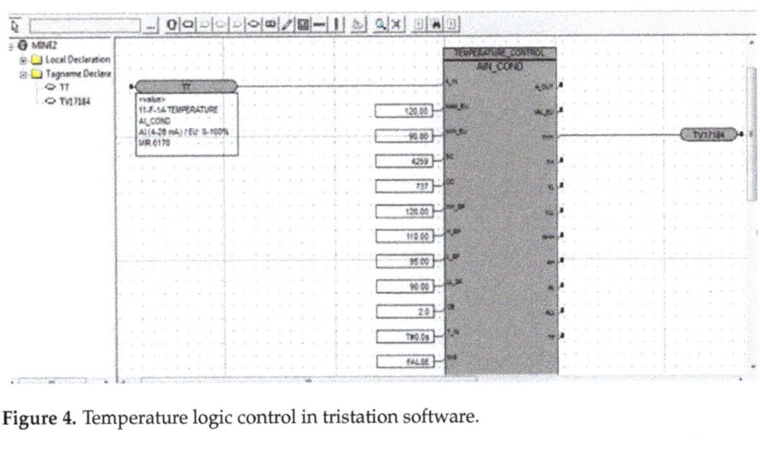

Figure 4. Temperature logic control in tristation software.

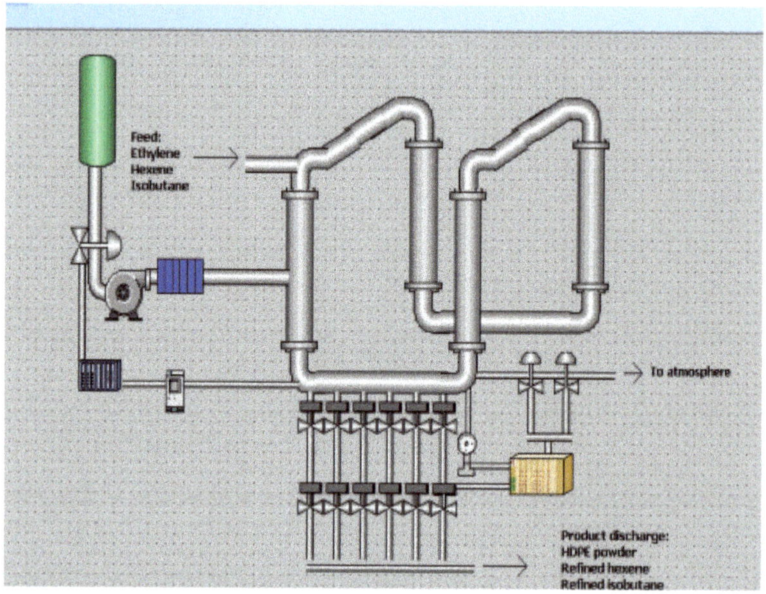

Figure 5. Build the HDPE reactor HMI.

3.4. Some Scenario Simulation

In this part we show some scenarios simulation as in Figure 6, when the temperature is increased the control system activate the cooling system to reduce the temperature. And so on in the case of high pressure or other parameters.

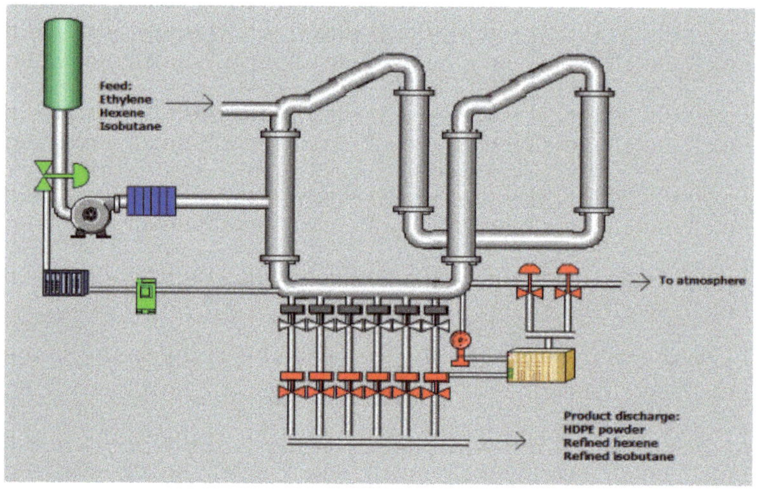

Figure 6. Scenario simulation High temperature.

4. Conclusions

In this paper, we have presented a system based theory that can deal with systemic failures when analyzing HDPE reactor. STPA has been applied to the lowest level of the HDPE plant. The methodology has been applied on the HDPE reactor showing that it can provide the same safety recommendations as other techniques (as HAZOP) but also considering other factors out of the scope of those techniques. The case study shows how STPA could replace or at least complement HAZOP as the hazard analysis technique for chemical and oil & gas industries. The advantage of using STPA lies in its systemic nature and its application to the whole socio-technical hierarchy. Another advantage of STPA is that it can give a potential recommendations to eliminate hazards using the same analysis (the closest with less change in variables- safe scenario shown in the table). At the end we have implemented this analysis in the control system to improve safety process in term of time, cost and accuracy i.e., reduce the human errors by displaying textual messages and corrective actions.

Author Contributions: Conceptualization, M.E.-A. and B.R.; methodology, M.E.-A.; software, A.A.; validation, M.E.-A., B.R.; formal analysis, M.E.-A.; investigation, M.E.-A.; resources, A.A.; data curation, M.E.-A.; writing—original draft preparation, M.E.-A. and B.R.; writing—review and editing, M.E.-A. and Z.Y.; visualization, B.R.; supervision; project administration, M.E.-A. and B.R; funding acquisition, M.E.-A. and B.R. All authors have read and agreed to the published version of the manuscript.

Funding: This research received no external funding.

Conflicts of Interest: The authors declare no conflict of interest.

References

1. Mechhoud, E.A.; Rouainia, M.; Rodriguez, M. A new tool for risk analysis and assessment in petrochemical plants. *Alex. Eng. J. (AEJ)* **2016**, *55*, 2919–2931. [CrossRef]
2. Leveson, N.G. *Engineering Safer World: Systems Thinking Applied to Safety*; MIT Press: Cambridge, MA, USA, 2011; ISBN 978-0-262-01662.
3. Yousefi, A.; Hernandez, M.R. A novel methodology to measure safety level of a process plant using a system theory based method (STAMP). *Process Saf. Environ. Prot.* **2020**, *136*, 296–309. [CrossRef]
4. Zhang, J.; Kim, H.; Liu, Y. Combining system-theortic process analysis and availability assessment: A subsea case study. *Proc. Inst. Mech. Eng. Part O J. Risk Reliab.* **2019**, *233*, 520–536.
5. Yousefi, A.; Hernandez, M.R. Using a system theory based method (STAMP) for hazard analysis in process indystry. *J. Loss Prev. Process Ind.* **2019**, *61*, 305–324. [CrossRef]
6. Rodríguez, M.; Díaz, I. A systematic and integral hazards analysis technique applied to the process industry. *J. Loss Prev. Process Ind.* **2016**, *43*, 721–729. [CrossRef]
7. Mechhoud, E.A.; Rouainia, M. Automated dependability analysis of a HDPE reactor. In Proceedings of the 2014 IEEE International Carnahan Conference on Security Technology (ICCST), Rome, Italy, 13–16 October 2014; pp. 1–5.
8. Mechhoud, E.A.; Rouaïnia, M.; Rodriguez, M. Functional modeling of a HDPE reactor using dhigraphs for process hazard analysis. In Proceedings of the 2016 International Conference on Modelling, Identification and Control (ICMIC), Algiers, Algeria, 15–17 November 2016; pp. 1–6.

engineering proceedings

MDPI

Proceeding Paper

ACO-Based Optimal MIMO Sliding Mode Controller Design for a New Reconfigurable Unmanned Aerial Vehicle †

Khedidja Bouhabza [1,*,‡], Mohamed Guiatni [1,‡], Yasser Bouzid [1,‡] and Mustapha Hamerlain [2,‡]

1 Complexe Systems Control and Simulators Laboratory, Ecole Militaire Polytechnique, Bordj el Bahri 16046, Algeria
2 Robotics Laboratory Head, Centre de Développement des Technologies Avancées, Algiers 16081, Algeria
* Correspondence: bouhabzakhadidja@gmail.com
† Presented at the 2nd International Conference on Computational Engineering and Intelligent Systems, Online, 18–20 November 2022.
‡ These authors contributed equally to this work.

Abstract: The Optimized Multiple Inputs and Multiple Outputs Sliding Mode controller (MIMO-SMC) is intended to control the reconfigurable UAV, which can take on a variety of configurations while maintaining the efficiency of a conventional quadrotor in terms of hovering and precision handling. The Ant Colony optimization algorithm is used to calculate controller gains and to ensure optimal performance, such as small errors and lower energy consumption. The control goal is to allow the proposed UAV to track its trajectory for various configurations in the face of uncertain parameters to impose the position, altitude, and yaw angle while stabilizing its roll and pitch angles. Finally, simulation results are used to demonstrate the efficiency and performance of the proposed controller, and we calculate the energy consumption for each configuration.

Keywords: optimized; MIMO sliding mode; reconfigurable UAV; configurations; quadrotor; ant colony optimization; trajectory; energy consumption

Citation: Bouhabza, K.; Guiatni, M Bouzid, Y.;Hamerlain, M. ACO-Based Optimal MIMO Sliding Mode Controller Design for a New Reconfigurable Unmanned Aerial Vehicle. *Eng. Proc.* **2023**, *29*, 9. https://doi.org/10.3390/engproc2023029009

Academic Editors: Abdelmadjid Recioui, Hamid Bentarzi and Fatma Zohra Dekhandji

Published: 11 January 2023

1. Introduction

Reconfigurable UAVs are a new type of UAV that can use adaptive morphology, which involves changing the shape of the vehicle during flight to accomplish a specific task in a unique environment [1]. Rotors/arm: tiltable, steerable, rotatable, reconfigurable, transformable, convertible, unconventional [2], multirotors: foldable [3], with rotating arms, with changeable dimensions, and so on.

The authors developed a transformable multirotor with two-dimensional multiple bonds in references [3,4]. The latter has a unique architecture that allows it to completely change its morphology.

Different approaches have been proposed in the literature for the different types of quadrotors such as PID (the Proportional-Integral-Derivative) in [5–9], the linear quadratic [10]; feedback linearization [11–13], fuzzy logic (FL) control [14], Backstepping [15,16], and control by sliding mode [17–19].

In [20,21] proposed a classical PID for a tilting body multirotor that can switch to a vertical configuration. Scientists used an adaptive PID controller to control a foldable quadrotor in reference [22].

Other authors created a multi-link foldable quadrotor based on the Linear Quadratic Integral (LQI) [2]. The author investigated the use of Linear Quadratic Regulation (LQR) on a foldable quadrotor in [3]. The authors proposed a nonlinear controller of a rotor quadrotor based on feedback linearization in [23]. In [24], an optimal gain scheduling backstepping controller based on a Particle Swarm Optimization (PSO) algorithm will be designed and tested for a Transformable Quadrotor. A new design with generic modeling and an adaptive backstepping controller of a transformable quadrotor can be found in [25].

The main contribution of this work is the optimal MIMO-SMC for the flight stability and rapid convergence of variable states of a new reconfigurable UAV in finite time. Furthermore, the MIMO-SMC's robustness will be tested against the uncertain parameters of the variable drone geometry. The primary control architecture is based on selecting between the various flight configurations. The gains of the MIMO-SMC controllers are obtained by using the Ant Colony Optimization (ACO) algorithm to minimize the error and energy criterion and establish which configuration consumed less energy.

The rest of the paper is structured as follows: Section 2 presents the reconfigurable UAV description and modeling. Section 3 explains the Integral Sliding Mode Controller (MIMO-SMC) design, and Section 4 describes the Ant Colony Optimization algorithm. Sections 5 and 6 present the simulation results as well as their interpretation. Section 7 examines a comparison of flight configurations. Section 8 is concerned with the conclusion and future works.

2. The Reconfigurable UAV Description and Modeling

The reconfigurable quadrotor has a very complex structure due to the change in configuration each time it adapts to a new morphology by slightly rotating the arms in [25].

2.1. *The Reconfigurable UAV's Description*

In this section, we will establish a reconfigurable UAV with four rotating arms that bend independently around the central body while keeping the angle of rotation $\alpha_i(t)|_{i=1,\ldots,4}$, using servomotors to allow for quick transitions between different configurations. To avoid propeller collisions, as shown in Figure 1a, six morphologies are used, as depicted in Figure 1b and Table 1.

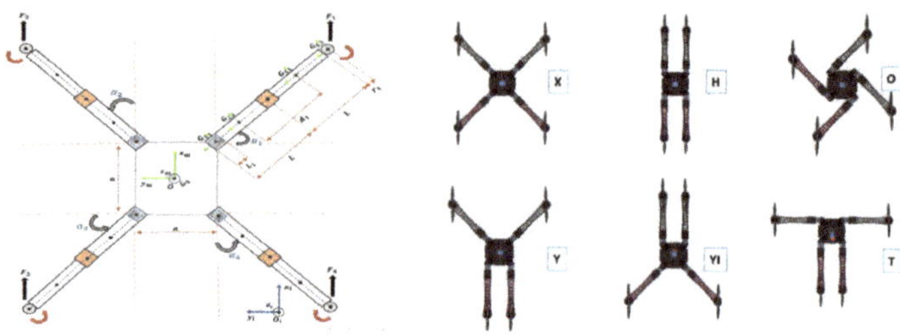

(a) (b)

Figure 1. (**a**): Description of the reconfigurable UAV [25], (**b**): the different configurations of reconfigurable UAV [25].

Table 1. Angles corresponding to each UAV's configuration.

Configuration	X	H	O	Y	YI	T
$\alpha_1(t)$	$\pi/4$	$\pi/2$	π	$\pi/4$	$\pi/2$	0
$\alpha_2(t)$	$\pi/4$	0	π	$\pi/4$	0	$\pi/2$
$\alpha_3(t)$	$\pi/4$	$\pi/2$	π	$\pi/2$	$\pi/4$	$\pi/2$
$\alpha_4(t)$	$\pi/4$	0	π	0	$\pi/4$	0

The angles of the arms can vary independently during each rotation of the arms, which means the quadrotor's configuration changes, the center of gravity moves, and the inertia matrix and roll and pitch moments, in this case, change based on the arm orientations.

To perform a specific task in a specific area, our quadrotor's configuration changes, it adopts an adaptive morphology to that task, and it can transform from the classic configuration "X" to other specific configurations by rotating or extending its arms.

2.2. The Reconfigurable UAV's Modeling

When the configuration changes, the global position of the CoG changes. As a result, the inertia matrix $\Im_{3\times3}(\alpha_i(t))$, roll $\tau_\varphi(\alpha_i(t))$ and pitch $\tau_\theta(\alpha_i(t))$, moments, in this case, vary based on the arm angles $\alpha_i(t)$. The linear and the angular velocity vectors of the body in the mobile frame $\Re_m$ are represented, respectively: $\Lambda^m = (u, v, w)^T \in \mathbb{R}^3$ and $\varsigma = (p, q, r) \in \mathbb{R}^3$. Let $Y = (\varphi, \theta, \psi)^T \in \mathbb{R}^3$ describe the orientation of the mobile and $\xi = (x, y, z)^T \in \mathbb{R}^3$ denote its position in the fixed frame $\Re_i$.

The relation between the velocities and the external forces $f^m = \left(f_x^m, f_y^m, f_z^m\right)^T \in \mathbb{R}^3$ and moments $\phi^m = \left(\phi_x^m, \phi_y^m, \phi_z^m\right)^T \in \mathbb{R}^3$ applied to CoG can be presented using the Newton–Euler formalism as:

$$\begin{bmatrix} m\Im_{3\times3}(\alpha_i(t)) & O_{3\times3} \\ O_{3\times3} & \Im_{3\times3}(\alpha_i(t))) \end{bmatrix}\begin{bmatrix} \dot{\Lambda}^m \\ \dot{\xi} \end{bmatrix} + \begin{bmatrix} \zeta \times m\Lambda^m \\ \zeta \times \Im(\alpha_i(t))\varsigma \end{bmatrix} = \begin{bmatrix} f^m \\ \phi^m \end{bmatrix} \tag{1}$$

2.3. Control Matrix

The control matrix $\Delta(\alpha_i(t)) \in \mathbb{R}^{4*4}$ transforms the angular speeds of the propellers $\Omega_i^2|_{i=1,\dots,4}$ to a total thrust force $\mathcal{T}$ and moments $\tau_\varphi, \tau_\theta, \tau_\psi$. It can be represented as follows:

$$\Delta = \begin{bmatrix} b & b[y_{4,1}(t) - y_G(t)] & b[x_G(t) - x_{4,1}(t)] & d \\ b & b[y_{4,2}(t) - y_G(t)] & b[x_G(t) - x_{4,2}(t)] & -d \\ b & b[y_{4,3}(t) - y_G(t)] & b[x_G(t) - x_{4,3}(t)] & d \\ b & b[y_{4,4}(t) - y_G(t)] & b[x_G(t) - x_{4,4}(t)] & -d \end{bmatrix}^T \tag{2}$$

where b and d are the thrust and drag coefficients, respectively, and $(x_{4,i}, y_{4,i})$ are the CoG coordinates of the rotors.

The relation between the total thrust force, the moments applied to CoG and propeller square velocities is expressed as follows:

$$\begin{bmatrix} \mathcal{T} \\ \tau_\varphi \\ \tau_\theta \\ \tau_\psi \end{bmatrix} = \Delta \begin{bmatrix} \Omega_1^2 \\ \Omega_2^2 \\ \Omega_3^2 \\ \Omega_4^2 \end{bmatrix} \tag{3}$$

The proposed UAV has twelve control inputs, where $u_1 = \mathcal{T}, u_2 = \tau_\varphi, u_3 = \tau_\theta, u_4 = \tau_\psi$ are used for the control of its altitude and attitude, while u_5, u_6, u_7, u_8 control the arms rotation. The model used to design the controller is given by Equation (4) as:

$$\begin{cases} \ddot{\varphi} = \beta_1(t)\dot{\theta}\dot{\psi} + \beta_2(t)\dot{\theta}\Omega_r + \beta_3(t)u_2 + \beta_4(t)\dot{\varphi}^2 \\ \ddot{\theta} = \beta_5(t)\dot{\varphi}\dot{\psi} + \beta_6(t)\dot{\varphi}\Omega_r + \beta_7(t)u_3 + \beta_8(t)\dot{\theta}^2 \\ \ddot{\psi} = \beta_9(t)\dot{\theta}\dot{\varphi} + \beta_{10}(t)u_4 + \beta_{11}(t)\dot{\psi}^2 \\ \ddot{x} = u_1\dfrac{u_x}{m_y} + \beta_{13}\dot{x} \\ \ddot{y} = u_1\dfrac{u_y}{m} + \beta_{14}\dot{y} \\ \ddot{z} = -g + u_1\dfrac{c_\varphi c_\theta}{m} + \beta_{12}\dot{z} \end{cases} \tag{4}$$

$$u_x = c_\psi s_\theta c_\varphi + s_\psi s_\varphi, u_y = s_\psi s_\theta c_\varphi - c_\psi s_\varphi \tag{5}$$

While $c(.) = \cos(.)$ and $s(.) = \sin(.)$.

Using Equations (1) and (4), the simplified control model can be determined as Equation (8). The state vector is given as:

$$X = [\varphi, \dot{\varphi}, \theta, \dot{\theta}, \psi, \dot{\psi}, z, \dot{z}, x, \dot{x}, y, \dot{y}]^T \tag{6}$$

such as:

$$X = [x_1, x_2, x_3, x_4, x_5, x_6, x_7, x_8, x_9, x_{10}, x_{11}, x_{12}]^T \tag{7}$$

$$\begin{cases} \dot{x}_1 = x_2 \\ \dot{x}_2 = \beta_1(t)x_4x_6 + \beta_2(t)x_4\Omega_r + \beta_3(t)u_2 + \beta_4(t){x_2}^2 \\ \dot{x}_3 = x_4 \\ \dot{x}_4 = \beta_5(t)x_2x_6 + \beta_6(t)x_2\Omega_r + \beta_7(t)u_3 + \beta_8(t){x_4}^2 \\ \dot{x}_5 = x_6 \\ \dot{x}_6 = \beta_9(t)x_2x_4 + \beta_{10}(t)u_4 + \beta_{11}(t){x_6}^2 \\ \dot{x}_7 = x_8 \\ \dot{x}_8 = -g + u_1\frac{\cos(x_1)\cos(x_3)}{m} + \beta_{12}x_8 \\ \dot{x}_9 = x_{10} \\ \dot{x}_{10} = u_1\frac{u_x}{m} + \beta_{13}x_{10} \\ \dot{x}_{11} = x_{12} \\ \dot{x}_{12} = u_1\frac{u_r}{m} + \beta_{14}x_{12} \end{cases} \tag{8}$$

where:

$$\begin{aligned} &\beta_1(t) = \frac{I_{yy}(\alpha_i(t), \mathrm{d}_i(t)) - I_{zz}(\alpha_i(t), \mathrm{d}_i(t))}{I_{xx}(\alpha_i(t), \mathrm{d}_i(t))}, \quad \beta_2(t) = \frac{-\mathfrak{J}_r}{I_{xx}(\alpha_i(t), \mathrm{d}_i(t))}, \beta_3(t) = \frac{1}{I_{xx}(\alpha_i(t), \mathrm{d}_i(t))} \\ &\beta_4(t) = \frac{-\mathcal{K}_{Ax}}{I_{xx}(\alpha_i(t), \mathrm{d}_i(t))}, \quad \beta_5(t) = \frac{I_{zz}(\alpha_i(t), \mathrm{d}_i(t)) - I_{xx}(\alpha_i(t), \mathrm{d}_i(t))}{I_{yy}(\alpha_i(t), \mathrm{d}_i(t))}, \\ &\beta_6(t) = \frac{\mathfrak{J}r}{I_{yy}(\alpha_i(t), \mathrm{d}_i(t))} \quad \beta_7(t) = \frac{1}{I_{yy}(\alpha_i(t), \mathrm{d}_i(t))}, \beta_8(t) = \frac{-\mathcal{K}_{Ay}}{I_{yy}(\alpha_i(t), \mathrm{d}_i(t))}, \\ &\beta_9(t) = \frac{I_{xx}(\alpha_i(t), \mathrm{d}_i(t)) - I_{yy}(\alpha_i(t), \mathrm{d}_i(t))}{I_{zz}(\alpha_i(t), \mathrm{d}_i(t))} \quad \beta_{10}(t) = \frac{1}{I_{zz}(\alpha_i(t), \mathrm{d}_i(t))}, \beta_{11}(t) = \frac{-\mathcal{K}_{Az}}{I_{zz}(\alpha_i(t), \mathrm{d}_i(t))}, \\ &\beta_{12} = \frac{-\mathcal{K}_{Dz}}{\mathrm{m}}, \beta_{13} = \frac{-\mathcal{K}_{Dx}}{\mathrm{m}}, \beta_{14} = \frac{-\mathcal{K}_{Dy}}{\mathrm{m}}, \quad \Omega_r = \sum_{i=1}^{4}(-1)^{i+1}\Omega_i^2 \end{aligned} \tag{9}$$

3. The Reconfigurable UAV's Control

3.1. Control Structure

The variable parameters of the reconfigurable drone are characterized by the change in its structure. Furthermore, the proposed MIMO-SMC will be applied and tested in the presence of these uncertain parameters, such as CoG, inertia, arm angles $\alpha_i(t)$ and control matrix, to ensure that the state variables converge rapidly towards the sliding surface and, ultimately, the equilibrium point in finite time.

The rotation of the servomotors close to the main body causes the arms to vary by angles α_i, whereas the rotation of the servomotors is attached to the primary arms γ_i. There are blocks that generate the desired trajectories (x_d, y_d, z_d and ψ). Our UAV's attitude and translations are controlled in two ways: by varying the position of the arms or the angular speeds of the four rotors. The two controls u_x and u_y are employed to calculate the desired roll and pitch angles. The servomotors that rotate the arms are controlled by PID controllers as follows:

3.2. MIMO Sliding Mode Control (MIMO-SMC) Design

This subsection describes the robust nonlinear controller design for a new reconfigurable UAV.

The sliding mode approach for MIMO nonlinear input-output systems has been proposed in [26]. To the best of our knowledge, this approach has never been applied for quadrotors. To do so, we follow the steps below:

- The state space model of UAV Equation (8) is written as generalized canonical;
- Sliding surfaces are chosen as Hurwitz polynomials;
- Sliding reachability conditions are selected;
- The highest-order derivatives of the control are obtained from the reachability condition.

We will go through the following procedures to synthesise the control law: We split our model state space of UAV Equation (8) into six subsystems, we write each subsystem in the form of differential input/output equations, we take the first subsystem

$$SS_1 = \begin{cases} \dot{x}_1 = x_2 \\ \dot{x}_2 = \beta_1(t)x_4x_6 + \beta_2(t)x_4\Omega_r + \beta_3(t)u_2 + \beta_4(t){x_2}^2 \end{cases} \quad (10)$$

The sliding surface is chosen as Hurwitz Polynomials, so we obtain the Sliding surface as follows:

$$S_i = \ddot{e}_i + \alpha_{j+1}\dot{e}_i + \alpha_j e_i \,; j = 1,3,5,7,9,11 \quad (11)$$

Using the candidate Lyapunov function:

$$V_i = \frac{1}{2}S_i^2 > 0 \quad (12)$$

$\dot{V}_i < 0$, which leads to $S_i\dot{S}_i < 0$, which is the necessary condition of sliding, is required to ensure stability in the Lyapunov sense. The following is how we decide on the attractiveness condition:

$$\dot{S}_i = -L_i \tanh(S_i) - M_i S_i = \dddot{e}_i + \alpha_{j+1}\ddot{e}_i + \alpha_j\dot{e}_i \quad (13)$$

As a result of inserting what we need in the previous Equation (13), the control law derivatives are:

$$\begin{aligned} \dot{U}_2 = \frac{1}{\beta_3}[\dddot{x}_{1d} - \beta_1(\beta_5x_2x_6 + \beta_6x_2\Omega_r + \beta_7U_3 + \beta_8{x_4}^2)(\beta_9(t)x_2x_4 + \beta_{10}(t)u_4 + \beta_{11}(t){x_6}^2) \\ -\beta_2(\beta_5x_2x_6 + \beta_6x_2\Omega_r + \beta_7U_3 + \beta_8{x_4}^2)\Omega_r - \alpha_2(\beta_1(t)x_4x_6 + \beta_2(t)x_4\Omega_r + \beta_3(t)u_2 + \beta_4(t){x_2}^2 \\ -\ddot{x}_{1d}) - \alpha_1(x_2 - \dot{x}_{1d}) - L_1\tanh(S_1) - M_1S_1 \end{aligned} \quad (14)$$

where α_1, α_2, L_1 and M_1 are positive parameters.

We repeat the previous steps to demonstrate asymptotic stability and extract the other controllers.

4. Optimization

We decided to employ optimization techniques because choosing parameters took a long time due to guesswork. We will propose a metaheuristics-based optimization technique and use it to identify the optimum control law parameters that were previously synthesized. Ant Colony Optimization (ACO) is used to obtain optimal parameters of the MIMO Sliding Mode Controller in [27,28].

For the creation of probable good solutions, ACO uses a pheromone matrix $\tau = \tau_{ij}$, and $\tau_0 > 0$ is the initial value of this matrix. The probability P_{ij} of selecting a node j at node i is given by Equation (15).

$$P_{ij} = \frac{\tau_{ij}(t)]^a[\eta_{ij}(t)]^b}{\sum_{i,j\in T^A}[\tau_{ij}(t)]^a[\eta_{ij}(t)]^b}, \; if \;\; i,j \in T^A \quad (15)$$

Heuristic functions are represented by:

$$\eta_{ij}(t) = \frac{1}{k_j} \quad (16)$$

a and b are constants that govern the pheromone and heuristic values' proportionate influence on the ant's decision. T^A is the path taken by the ant A. The amount of pheromone $\Delta\tau_{ij}^A$ on each path can be calculated as follows:

$$\Delta\tau_{ij}^A = \begin{cases} \frac{L^{min}}{L^A}\ if\ \ i,j \in T^A \\ 0 \end{cases} \quad (17)$$

L^A denotes the value of the A ant's objective function. L^{min} is the best response that the ants have established up to the present iteration. Pheromone dissipation is a method of preventing an infinite rise in real ants. It also enables the forgetting of negative decisions.

$$\tau_{ij}(t) = P\tau_{ij}(t-1) + \sum_{A=1}^{NA} \Delta\tau_{ij}(t) \quad (18)$$

where NA is the number of ants, and P is the evaporation rate $0 < p <= 1$. Using the optimization approach mentioned in the previous paragraph, we explore optimization in the situation of search parameter's optimal control laws applied to the quadrotor. To do so, we use the control laws from section three and modify them.

The number of iterations is set at 100, and we take the performance criteria for optimizing the energy supplied by the controller J_1 and the mean quadratic error J_2, where:

$$\begin{cases} J_1 = \frac{1}{k}\sum_{i=1}^{4} u_i^2 \\ J_2 = \frac{1}{k}\Sigma e_i^2 \quad\quad with \quad i = \varphi, \theta, \psi, z,, x, y. \end{cases} \quad (19)$$

5. Simulation Results

We will perform a flight scenario to assess the efficiency of the proposed controller for reconfigurable UAV attitude and altitude stabilization, which are implemented and discussed in this section. We used the Simulink/Matlab code to create a trajectory scenario (square). Our proposed UAV will follow each trajectory with different morphologies (Table 1). Because of the figures' large number, we chose just two configurations. The first will be in the form "O", and the second will be in the form "YI", as shown in Figures 2–5.

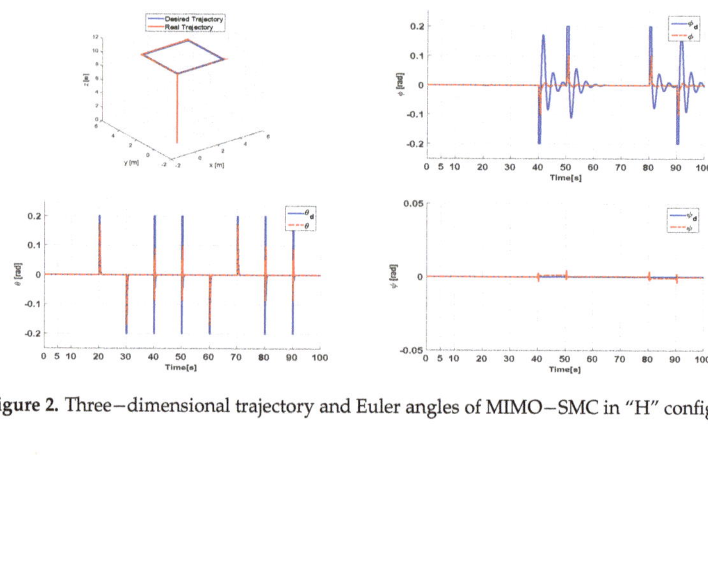

Figure 2. Three−dimensional trajectory and Euler angles of MIMO−SMC in "H" configuration.

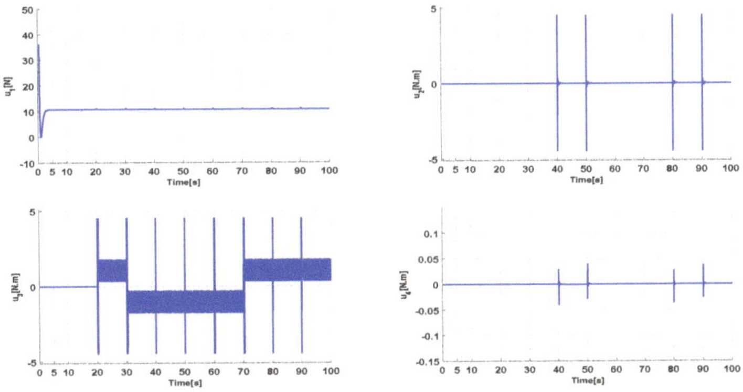

Figure 3. Input signalsof MIMO−SMC in "H" configuration.

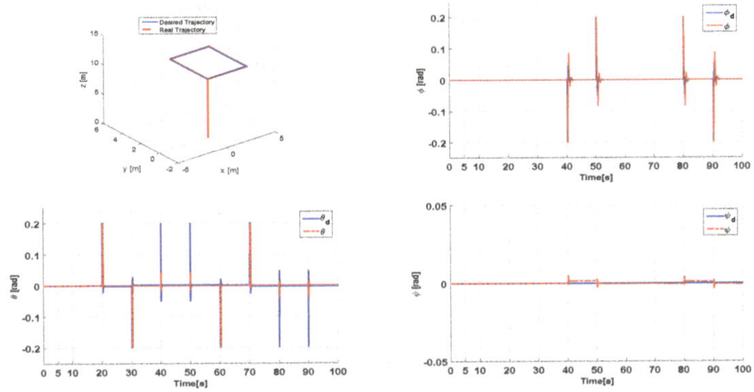

Figure 4. Three−dimensional trajectory and the responses positions of MIMO−SMC in "YI" configuration.

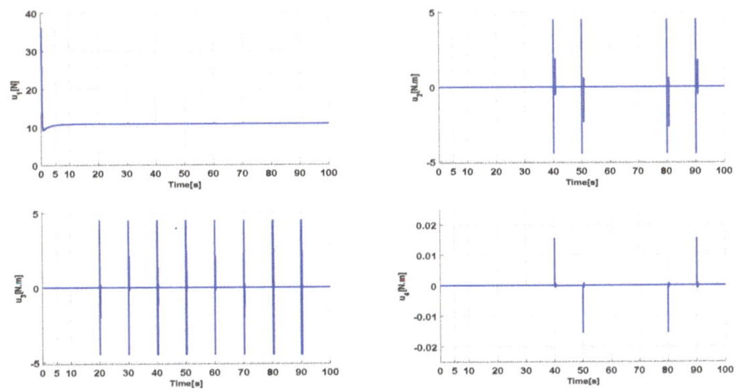

Figure 5. Input signals of MIMO−SMC in "YI" configuration.

6. Results Interpretation

Figures 2 and 4 show that the reconfiguration can follow the desired trajectory for two scenarios and for the two configurations. We observe some errors along the three axes at first (x, y, z) for the "H" configuration compared to the "YI" configuration. The system's responses for MIMO-SMC show the rapid convergence of variable states (x, y, z)

to their reference states (x_d, y_d, z_d) in finite time. We also observe a good tracking of Euler angles (ϕ, θ, ψ) to their desired attitude $(\phi_d, \theta_d, \psi_d)$. We can also notice that once the desired positions and attitudes are reached, the errors in position and attitude converge to zero.

Figures 3 and 5 show control inputs, with the control signal u_1 stabilizing around a constant that corresponds to the altitude z and the sliding surface S_z. The changes in the two control signals u_2 and u_3 are determined by the dynamics of $S\phi$ and $S\theta$ and u_3 in the square trajectory, and for the "H" configuration, it is more energetic compared to the "YI" configuration, as well as very low values of u_4, resulting in low power consumption.

We remark that the "YI" configuration consumes less energy compared to the "H" configuration. The proposed control technique ensures rapidity, accuracy and strong stability of the system, demonstrating that the quadrotor can track its desired trajectory in our scenario and for two configurations of our reconfigurable UAV. Furthermore, it is obvious that MIMO-SMC decreases the chattering phenomena, errors and energy consumption.

7. Comparative Study

First, let us define the energy consumed by the vehicle during the flight.

$$E_c = \int_{t_0}^{t_f} \sum_{i=1}^{4} \tau_i(t)\Omega_i(t)dt \tag{20}$$

where $\tau_i(t)$ is the torque generated by motor i, and $\Omega_i(t)$ is the angular speeds of the propellers at time t. We assess the consumed energy by the reconfigurable UAV in each configuration for a trajectory (square) under an optimal control approach (MIMO-SMC based on ACO) using Equation (20), which is summarized in Table 2.

Table 2. Consumed energy by each configuration.

Morphology	X	H	O	Y	YI	T
$E_c(KJ)Square$	5.6639	5.6615	5.6687	5.8804	**5.4341**	6.6764

We calculate the percentage difference between the consumed energy obtained with the classic "X" configuration compared to the energy consumed with the YI configuration, which is 3%, and between the YI configuration and H, it is 18%. The YI configuration consumes less energy compared to other special configurations ("X", "H", O", "Y" and "T"). This comparative study quantifies the amount of energy consumed by the reconfigurable UAV in each configuration. To optimize the energy consumed along a predefined trajectory, we change the configuration "X", "H", "O", "Y", "YI" and "T" of the reconfigurable drone during the flight to accomplish a well-defined mission, which will be the objective of future work.

8. Conclusions and Future Work

We are interested in reconfigurable UAVs in this work. When compared to conventional drones, there are several advantages, including the reduction in energy consumed during flight, handling and transporting of objects, execution of specific missions, and various configurations. These new drones are distinguished primarily by a transformable and complex mechanical structure, which makes modeling and control of the latter very difficult tasks. We were able to establish the use of a generic model in the literature in this manuscript. After that, we ordered the drone proposed by the Multiple Inputs and Multiple Outputs Sliding Mode Controller in order to stabilize it in flight while taking into account uncertain parameters caused by aerial transformation for our scenario and for each configuration. As well as the adequate gains of the control, we used the Ant Colony optimization technique. Finally, several simulations were run to demonstrate the efficiency of the proposed controller in terms of precision, speed, and stability, and we concluded which configuration consumed the least amount of energy.

We hope to confirm the results obtained in the future by applying the proposed controller to a real reconfigurable UAV, estimating disturbances, using genetic algorithms, and changing configurations during flight.

Author Contributions: Methodology, M.G.; software, K.B., M.G., Y.B., M.H.; validation, K.B., M.G., Y.B., M.H.; formal analysis, K.B., M.G., Y.B.; writing—original draft preparation, K.B.; writing—review and editing, K.B.; visualization, M.G.; funding acquisition, M.G., Y.B. All authors contributed equally to this work. All authors have read and agreed to the published version of the manuscript.

Funding: This research received no external funding

Institutional Review Board Statement: Not applicable

Informed Consent Statement: All authors have read and agreed to publish this work.

Data Availability Statement: Not applicable

Conflicts of Interest: The authors declare that they have no conflict of interest.

References

1. Falanga, D.; Kleber, K.; Mintchev, S.; Floreano, D.; Scaramuzza, D. The Foldable Drone: A Morphing Quadrotor That Can Squeeze and Fly. *IEEE Robot. Autom. Lett.* **2019**, *4*, 209–216. [CrossRef]
2. Zhao, M.; Kawasaki, K.; Chen, X.; Noda, S.; Okada, K.; Inaba, M. Whole-body aerial manipulation by transformable multirotor with two-dimensional multilinks. In Proceedings of the 2017 IEEE International Conference on Robotics and Automation (ICRA), Singapore, 29 May–3 June 2017; pp. 5175–5182.
3. Wallace, D.A. Dynamics and Control of a Quadrotor with Active Geometric Morphing. Master's Thesis, University of Washington, Seattle, WA, USA, 2016.
4. Zhao, M.; Anzai, T.; Shi, F.; Chen, X.; Okada, K.; Inaba, M. Design, modeling, and control of an aerial robot dragon: A dual-rotor-embedded multilink robot with the ability of multi-degree-of-freedom aerial transformation. *IEEE Robot. Autom. Lett.* **2018**, *3*, 1176–1183. [CrossRef]
5. Bouabdallah, S.; Noth, A.; Siegwart, R. PID vs. LQ control techniques applied to an indoor micro quadrotor. In Proceedings of the 2004 IEEE/RSJ International Conference on Intelligent Robots and Systems (IROS)(IEEE Cat. No. 04CH37566), Sendai, Japan, 28 September–2 October 2004; Volume 3, pp. 2451–2456.
6. Bressan, G.; Invernizzi, D.; Panza, S.; Lovera, L. Attitude control of multirotor uavs: Cascade p/pid vs pi-like architecture. In Proceedings of the 5th CEAS Specialist Conference on Guidance, Navigation and Control-EuroGNC, Milano, Italy, 3–5 April 2019; pp. 1–20.
7. Atheer, L.S.; Haider, A.M.; Khalaf, S.G. Flight PID controller design for a UAV quadrotor. *Sci. Res. Essays* **2010**, *5*, 3660–3667.
8. Bouabdallah, S. *Design and Control of Quadrotors with Application to Autonomous Flying;* Technical Report; Epfl: Lausanne, Switzerland, 2007.
9. Reizenstein, A. Position and Trajectory Control of a Quadcopter Using PID and LQ Controllers. Master's Thesis, Linköping University, Linköping, Sweden, 2017.
10. Araar, O.; Aouf, N. Full linear control of a quadrotor UAV, LQ vs. H∞. In Proceedings of the 2014 UKACC International Conference on Control (CONTROL), Loughborough, UK, 9–11 July 2014; pp. 133–138.
11. Choi, Y.; Ahn, H. Nonlinear control of quadrotor for point tracking: Actual implementation and experimental tests. *IEEE/ASME Trans. Mechatronics* **2014**, *20*, 1179–1192. [CrossRef]
12. Mukherjee, P.; Waslander, S.L. Direct adaptive feedback linearization for quadrotor control. In Proceedings of the AIAA Guidance, Navigation, and Control Conference, Minneapolis, MN, USA, 3–16 August 2012; p. 4917.
13. Voos, H. Nonlinear control of a quadrotor micro-UAV using feedback-linearization. In Proceedings of the 2009 IEEE International Conference on Mechatronics, Malaga, Spain, 14–17 April 2009; pp. 1–6.
14. Deepak, G.; Cheolkeun, H. Control of a Quadrotor Using a Smart Self-Tuning Fuzzy PID Controller. *Int. J. Adv. Robot. Syst.* **2013**, *10*, 380.
15. Huo, X.; Huo, M.; Karimi, H.R. Attitude stabilization control of a quadrotor UAV by using backstepping approach. *Math. Probl. Eng.* **2014**, *2014*, 749803. [CrossRef]
16. Kobilarov, M. Trajectory control of a class of articulated aerial robots. In Proceedings of the 2013 International Conference on Unmanned Aircraft Systems (ICUAS), Atlanta, GA, USA, 28–31 May 2013; pp. 958–965.
17. Utkin, V. Variable structure systems with sliding modes. *IEEE Trans. Autom. Control* **1977**, *22*, 212–222. [CrossRef]
18. Slotine, J.; Li, W. *Applied Nonlinear Control;* Prentice Hall: Englewood Cliffs, NJ, USA, 1991; Volume 199.
19. Muño, F.; González-Hernández, I.; Salazar, S.; Espinoza, E.S.; Lozano, R. Second order sliding mode controllers for altitude control of a quadrotor UAS: Real-time implementation in outdoor environments. *Neurocomputing* **2017**, *233*, 61–71. [CrossRef]
20. Elfeky, M.; Elshafei, M.; Saif, A.W.A.; Al-Malki, M.F. Modeling and simulation of quadrotor UAV with tilting rotors. *Int. J. Control. Autom. Syst.* **2016**, *14*, 1047–1055. [CrossRef]

21. Hintz, C.; Torno, C.; Carrillo, L.R.G. Design and dynamic modeling of a rotary wing aircraft with morphing capabilities. In Proceedings of the 2014 International Conference on Unmanned Aircraft Systems (ICUAS), Orlando, FL, USA, 27–30 May 2014; pp. 492–498.
22. Riviere, V.; Manecy, A.; Viollet, S. Agile robotic fliers: A morphing-based approach. *Soft Robot.* **2018**, *5*, 541–553. [CrossRef] [PubMed]
23. Nemati, A.; Kumar, M. Non-linear control of tilting-quadcopter using feedback linearization based motion control. In Proceedings of the Dynamic Systems and Control Conference, San Antonio, TX, USA, 22–24 October 2014; American Society of Mechanical Engineers: New York, NY, USA, 2014; Volume 46209, p. V003T48A005.
24. Derrouaoui, S.H.; Bouzid, Y.; Guiatni, M. PSO Based Optimal Gain Scheduling Backstepping Flight Controller Design for a Transformable Quadrotor. *J. Intell. Robot. Syst.* **2021**, *102*, 1–25. [CrossRef]
25. Derrouaoui, S.H.; Bouzid, Y.; Guiatni, M. Towards a new design with generic modeling and adaptive control of a transformable quadrotor. *Aeronaut. J.* **2021**, *125*, 2169–2199. [CrossRef]
26. Lu, X.Y.; Spurgeon, S.K. Asymptotic stabilisation of multiple input nonlinear systems via sliding modes. *Dyn. Control* **1998**, *8*, 231–254. [CrossRef]
27. Colorni, A.; Dorigo, M.; Maniezzo, V. Distributed optimization by ant colonies. In Proceedings of the 1st European Conference on Artificial Life, Paris, France, 11–13 December 1991; pp. 134–142.
28. Dorigo, M. Optimization Learning and Natural Algorithms. Ph.D. Thesis, Politecnico di Milano, Milan, Italy, 1992.

engineering proceedings

MDPI

Proceeding Paper

Multiobjective Optimization of the Performance of Safety Systems †

Boukrouma Houcem Eddine [1,*], Bendib Riad [2], Zennir Youcef [2] and Mechhoud El-Arkam [2]

1 Research Laboratory on Surface and Interface Physics and Chemistry LRPCSI, Université 20 Août 1955 Skikda, Skikda 21000, Algeria
2 Automatic Laboratory of Skikda, Université 20 Août 1955 Skikda, Skikda 21000, Algeria
* Correspondence: hd.boukrouma@univ-skikda.dz
† Presented at the 2nd International Conference on Computational Engineering and Intelligent Systems, Online, 18–20 November 2022.

Abstract: The activation of the safety-instrumented systems in industrial processes is carried out after the occurrence of specific deviations (dangerous situations) from normal operation (normal situations), but in some cases, the safety-instrumented systems are activated in the absence of deviations or requests; these are the unwanted activations. The system chosen in this study is a system with high-pressure gas and inflammable gas, and it is protected by a firefighting system that prevents any kind of accident in order to protect humans, systems, and the environment. The activation of the emergency shutdown system causes stoppage of the whole system by closing the input and output valves. This paper presents the optimization of the voting redundancies of safety-instrumented systems by a multiobjective genetic algorithm. The objectives to optimize are the average probability of dangerous failure on demand, which represents the system safety integrity, and the spurious trip rate, which presents the activation of a safety function without the presence of a demand.

Keywords: optimize; vote; redundancy; safety instrumented system; emergency shutdown system; genetic algorithm

Citation: Eddine, B.H.; Riad, B.; Youcef, Z.; El-Arkam, M. Multiobjective Optimization of the Performance of Safety Systems. *Eng. Proc.* **2023**, *29*, 10. https://doi.org/10.3390/engproc2023029010

Academic Editors: Abdelmadjid Recioui, Hamid Bentarzi and Fatma Zohra Dekhandji

Published: 13 January 2023

1. Introduction

In most real-world problems, several goals must be satisfied simultaneously in order to obtain an optimal solution. The search for the best compromise between availability and costs (operational and life cycle) is at the heart of the industry's concerns.

A firefighting system or FF is a system that is used in hazardous areas to prevent situations that could have catastrophic effects economically, environmentally, or operationally. They are designed to minimize the consequences of emergencies, such as injury to personnel or damage to equipment [1]. In this context, the firefighting system not only serves to protect the installation in case of deviation, but also to protect the people in the vicinity.

This study was devoted to the optimization of the performances of FF of a combined cycle power system by the minimization of the rate of spurious trip (STR) as well as the probability failure on demand (PFD). This is done by acting on the vote of the architecture (KooN) of this system.

2. Abbreviation

CCF	Common cause failure	Avg	average
β	Factor for quantification of C	ESD	Emergency shutdown
βD = βDD	β for dangerous detected (DD) failures	FE	Final Element
βSD	β for safe detected (SD) failures	KooN	K out of N
βSU	β for safe undetected (SU) failures	LS	Logic Solver
λD	Dangerous failure rate	MRTS	Mean repair time for SU failures
λDD	Dangerous detected failure rate	MTTRSD	Mean time to restoration for SD failures
λDU	Dangerous undetected failure rate	S	Sensors
λS	Safe failure rate	SD	Safe Detected Failures
λSD	Safe detected failure rate	SIF	Safety Instrumented Function
λSU	Safe undetected failure rate	SIS	Safety Instrumented System
λDind	Dangerous detected independent failure rate	STR(KooN)	STR for KooN architecture
λSind	Safe detected independent failure rate	STR	Spurious trip rate
		SU	Safe Undetected Failures
		T1	Interval between tests

3. Identify the System

The gas turbine-based power plant is characterized by its relatively low capital cost compared with the steam power plant. One of the technologies adopted nowadays for efficiency improvement is the combined cycle. Combined cycle technology is now well established and offers superior efficiency to any of the competing gas turbine-based systems that are likely to be available in the medium term for large-scale power generation applications. A combined-cycle power system (Figure 1) typically uses a gas turbine to drive an electrical generator, and recovers waste heat from the turbine exhaust to generate steam. The steam from waste heat is run through a steam turbine to provide supplemental electricity [2]. The overall electrical efficiency of a combined-cycle power system is typically in the range of 50–60%—a substantial improvement over the efficiency of a simple, open-cycle application of approximately 33%.

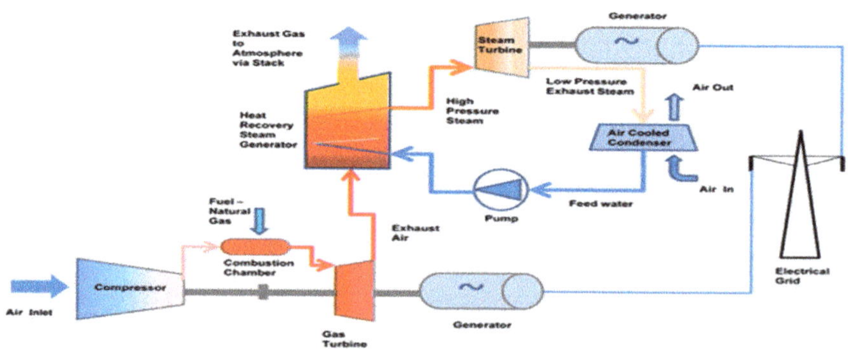

Figure 1. Block diagram of combined cycle power plan.

4. Firefighting System

The function of fire protection system is to provide safety (in terms of lives) from fire hazards and to provide protection (in terms of property) from fire hazards. Additionally, the fire protection system must be able to provide early fire warnings and fire suppression. Fire protection systems are designed in accordance with the requirements of NFPA and other standards [3].

The FF system contains three subsystem sensors that send the data from the site or the machine equipped with these sensors to the logic solver that treats the output of sensors and makes the decision regarding the intervention of the extinguishing system (Table 1).

Table 1. Elements constituting the FF system with the presentation of the architecture (KooN type) of each of these elements.

Elements	Architecture
Sensors	1oo3
Logic solver	1oo3
Extinguishing system	1oo3

In addition to failures on demand that affect the availability of the system when we need its safety action, the FF system can be activated in the absence of deviations by false signals from its sensors, an incorrect decision of the logic solver, an incorrect action of the logic unit, or an unwanted action. This activation is characterized by a spurious activation rate.

5. The Problem to Be Optimized

The firefighting system serves not only to protect the installation in case of deviation but also to protect the man in the surroundings and ensure the generation of energy necessary for many facilities. This system starts since the date (test phase) of their putting into service of the untimely activations of the FF system without the presence of a real demand of the installation, and this causes the shutdown of the entire installation, the loss of production, and loss of time to resume the work.

The optimization of FF performances will allow a reduction in the frequency of failures of this system and a reduction in the unwanted activations of this system to avoid such problems as the loss of production and wasted production time.

6. Classification of Firefighting Failures

Some failures of safety systems can lead to dangerous situations, and others can lead to false activations (but without any danger), so the failures of our safety system (FF) can be classified according to their effects in two categories [4].

(1) Dangerous failures: A failure that has the potential to put the safety-related system in a dangerous state or prevent a safety function from operating when required (demand mode). For these reasons, it is expressed with the probability of failure on demand (PFD). PFD is a measure of the effectiveness of a safety function. It expresses the likelihood that the safety function does not work when required to.

(2) Safe failures or spurious activations: These are failures that do not have the potential to put the safety-related system in a dangerous state or make it unable to perform its function [5].

The term "activation" indicates that there is some transition from one state to another and the term "spurious" indicates that the causes of the triggering are false, incorrect, and unreal [1].

Spurious activations of firefighting system can cause partial or complete shutdowns of the facilities, so it is necessary to reduce its occurrence to

(1) avoid production losses due to shutdowns, and
(2) avoid the risks that may appear during the restart phase.

The spurious activation rate or spurious trip rate is defined as the average number of spurious activations of a safety function per unit of time [6].

7. Redundancy of Systems

In order to reduce the probability that a safety system does not fulfill its security function at the moment it is requested, a solution consists of redundant (totally or partially) certain elements constituting that system (sensors, logic unit, terminal elements, and even transmission means). Note that the redundancy can be realized with identical materials or with different technologies.

A redundancy of 1ooN is the best to ensure the safety function, but it influences negatively the rate of spurious activations because with this architecture it is enough of one element to activate spuriously to cause the activation of all the system [7].

The optimization of the spurious activations and failure on demand will allow us to minimize the number of spurious activations, minimize the lost time to restart the system and the costs of stopping production, and minimize the dangerous situation in which the safety activity is required.

A problem such as this is very complicated, but we can solve it with analytical methods; in fact, the use of metaheuristic methods such as the genetic algorithm (GA) is the best way to solve this problem.

8. Optimization (STR) and PFD

The STR and PDF of a well-defined safety function provided by a given SIS is determined by calculating and combining the same variables of its three subsystems (S, LS, and FE). This can be expressed by the following general formula:

〖PFD 〗_avg = 〖PFD 〗_avg (S) + 〖PFD 〗_avg (LS) + 〖PFD 〗_avg (FE).

STRavgSIS = STRavg (S) + STRavg (LS) + STRavg (FE).

Obviously, each of these three subsystems is represented by a KooN architecture [2]. Reference [3] explains the formula of STR(KooN) and PDF (KooN):

$$\text{PFDmoy (KooN)} = A_N^{N-K+1}\lambda_{\text{Dind}}^{N-K+1}\prod_{i=1}^{N-K+1}\text{MDT}_{\text{i001}}+\beta\,\lambda\text{DU}\,(\frac{\text{T1}}{2}+\text{MRT})+\beta\text{D}\,\lambda\text{DD MTTR}.$$

$$\text{STR(KooN)} = A_N^K\,\lambda_{\text{Sind}}^K\prod_{i=1}^{K-1}\text{MDTS}_{\text{i001}}+\beta\,\lambda\text{SU}+\beta\text{D}\,\lambda\text{SD}. \quad (1)$$

with

$$A_N^{N-K+1} = \frac{N!}{(K-1)!}$$

$$\lambda\text{D} = \lambda\text{DD} + \lambda\text{DU}$$

$$\lambda\text{Dind} = (1-\beta)\,\lambda\,\text{DU} + (1-\beta\,\text{D})\,\lambda\text{DD}$$

$$\lambda\text{S} = \lambda\text{SD} + \lambda\text{SU}$$

$$\lambda\text{Sind} = (1-\beta\,\text{SU})\times\lambda\,\text{SU} + (1-\beta\,\text{SD})\,\lambda\text{SD}$$

$$〖\text{MDT} 〗\_\text{i001} = \lambda\text{DU}/\lambda\text{D}(\text{T1}/2+\text{MRT})+\lambda\text{DD}/\lambda\text{D.MTTR} \quad (2)$$

$$〖\text{MDTS} 〗\_\text{i001} = \lambda\text{SU}/\lambda\text{S}(\text{T1}/2+\text{MRTS})+\lambda\text{SD}/\lambda\text{S.MTTRS}. \quad (3)$$

We act on the voting of each subsystem because the system is in the production phase, which means we cannot modify the number of elements (N), so we can only act on the voting (KooN) of each subsystem sensor subsystem, logic solver subsystem, and final element subsystem. Multiobjective optimization is an area of multiple criteria decision making that is concerned with mathematical optimization problems involving more than one objective function to be optimized simultaneously. In our study, the objectives we try to optimize are the PFD and the STR of an emergency shutdown system.

9. Genetic Algorithm

A genetic algorithm is a search heuristic that is inspired by Charles Darwin's theory of natural evolution. This algorithm reflects the process of natural selection by which the fittest individuals are selected for reproduction in order to produce offspring of the next generation. All of this has been done via the following steps of the GA [8].

Notion of Natural Selection

The process of natural selection starts with the selection of fittest individuals from a population. They produce offspring which inherit the characteristics of the parents and will be added to the next generation. If parents have better fitness, their offspring will be better than parents and have a better chance of survival. This process keeps on iterating, and, at the end, a generation with the fittest individuals will be found. This notion can be applied for a search problem. We consider a set of solutions for a problem and select the best set [9].

At first, the coding to be used must be defined. Then, by using a random process, an initial population of strings is created. Next, a set of operators is used to take this initial population to generate successive populations, which hopefully improve with time. The main operators of the genetic algorithms are reproduction, crossover, and mutation.

Reproduction is a process based on the objective function (fitness function) of each string. This objective function identifies how good a string is. Thus, strings with higher fitness value have a larger probability of contributing offspring to the next generation.

Crossover is a process by which members of the last population are mated at random in the mating pool. A pair of offspring is generated, combining elements from two parents (members), which hopefully have improved fitness values. Mutation is the occasional (with small probability) random alteration of the value of a string position. In fact, mutation is a random-walk process through the coded parameter space. Its purpose is to ensure that important information contained within strings is not lost prematurely.

10. Results and Discussion

In order to facilitate the use of genetic algorithms, their execution is now fully supported by the fully supported by the Optimization Toolbox of the MATLAB environment. The gene code represented in Table 2.

Table 2. The gene of the code used in the GA.

K_S	$T1_S$	K_{LS}	$T1_{LS}$	K_{FE}	$T1_{FE}$

The variables in our code are the voting KooN of the sensors and logic solver but the final elements are two valves working together when the logic solver send a signal of closing so the voting of the final element is 2oo2. Moreover, the other variable is the T1 of each element of the FF system.

After executing the GA to optimize the STR of the ESD based on choosing the best voting and the best time between periodic tests, the results of the optimization can be presented, as in Table 3 [10].

Table 3. Results of the optimization the STR of FF system.

Variables			**Objective Functions**	
K1 (Sensors)	**K2 (UL)**	**K3 (Extg Sys)**	**PFDavg**	**STR**
2	2	2	0.00878	1.0257×10^{-5}
2	1	1	0.00855	3.5591×10^{-5}
2	1	2	0.00863	3.5516×10^{-5}

To get a PFD = 0.00878 and STR = 1.0257 $\times$ 10^{-5}, the best voting for the sensors is 2oo3 with a voting of logic solver 2oo3—the same voting for the extinguishing system. To get a PFD = 0.00855 and STR = 3.5591 $\times$ 10^{-5}, the best voting for sensors is 2oo3. For the logic solver, 1oo3 is the same for the extinguishing system and to get PFD = 0.00863. To get an STR = 3.5516 $\times$ 10^{-5}, the best voting for sensors is 2oo3, and for the logic solver 1oo3 and 2oo3 are the best voting for the extinguishing system.

11. Conclusions

As a conclusion to our work, it should be noted that spurious activation must be taken into consideration in the performance evaluation of instrumented safety systems due to the important economic losses caused by this type of failure.

For the optimization of the performance of a safety system, the use of genetic algorithms is a powerful tool to help in the decision-making process for the choice of an adequate architecture.

The GA still has some limitations, such as the coding step, which requires time and good knowledge of the system; in addition it relies on hazards in its operations.

That is why integrating machine learning will be more useful in solving optimization problems to provide methods of optimization with a higher, intelligent rate.

Author Contributions: Conceptualization, B.H.E. and B.R. and Z.Y.; methodology, B.H.E.; software, B.R.; Validation, B.R. and Z.Y. and M.E.-A.; formal analysis, B.H.E. and B.R.; writing original draft preparation, B.R.; Supervision review and editing, Z.Y.; project administration, M.E.-A. laboratory supervision. All authors have read and agreed to the published version of the manuscript.

Funding: This research received no external funding.

Data Availability Statement: Not applicable.

Conflicts of Interest: The authors declare no conflict of interest.

References

1. Fatima, M.; Assia, H.; Habib, H. Commande d'un système éolien basé sur une MADA. *IJSET* **2019**, *11*, 9–14.
2. *Design and Operation of Heat Exchangers and Their Networks*; Roetzel, W.; Luo, X.; Chen, D. (Eds.) Elsevier: Amsterdam, The Netherlands, 2020; ISBN 978-0-12-817894-2.
3. Combined Cycle Manual by Hyundai. 2020. Available online: https://www.hyundai.com/content/dam/hyundai/in/en/data/connect-to-service/owners-manual/kona.pdf (accessed on 1 November 2022).
4. *Standard IEC 16508*; Functional Safety of Safety-Related Electrical/Electronic/Programmable Electronic Systems. International Electrotechnical Commission: Geneva, Switzerland, 2000; Parts 1 to 7.
5. Lundteigen, M.A.; Rausand, M. *Spurious Activation of Safety Instrumented Systems in the Oil and Gas Industry: Basic Concepts and Formulas*; Elsevier: Amsterdam, The Netherlands, 2007.
6. Bae, J.-H.; Shin, S.-C.; Park, B.-C.; Kim, S.-Y. Design optimization of ESD (Emergency shutdown) System for offshore process based on reliability analysis. In Proceedings of the 2016 International Conference on Design Engineering and Science (ICDES 2016), Kuala Lumpur, Malaysia, 27–29 February 2016.
7. Thuy, L.; Adjadj, A.; Chaumette, S.; Bouchet, S.; de Dianous, V. Performance Evaluation of Engineered Safety Barriers. 2008. Available online: https://www.mementodumaire.net/wp-content/uploads/2012/03/Omega_10_Evaluation_BTS_v2_0908_web.pdf (accessed on 1 November 2022).
8. Innal, F.; Dutuit, Y.; Chebila, M. Reliability Engineering and System Safety: Safety and operational integrity evaluation and design optimization of safety instrumented systems. *Reliab. Eng. Syst. Saf.* **2015**, *134*, 32–50. [CrossRef]
9. Barros, D.B. Natural Selection as a Mechanism. *Philos. Sci.* **2008**, *75*, 306–322. [CrossRef]
10. *MATLAB 2017 Software*; MathWorks, Inc.: Natick, MA, USA, 2017.

engineering proceedings

MDPI

Proceeding Paper

Short-Term Wind Forecasting in Adrar, Algeria, Using a Combined System †

Farouk Chellali [1,2]

[1] Signals and Systems Laboratory, Institute of Electric and Electrinic Engineering, University of Boumerdes, Boumerdes 35000, Algeria; f.chellali@gmail.com
[2] Department of Mechanical Engineering, University of Djelfa, Djelfa 17000, Algeria

† Presented at the 2nd International Conference on Computational Engineering and Intelligent Systems, Online, 18–20 November 2022.

Abstract: Forecasting wind speed using only one single model may not be satisfactory; however, the combination of several models may provide better results, especially in case of doubt about the best estimator (predicator). In this paper, we propose a short-term forecasting system in order predict the hourly averaged wind speed (HAWS) in the region of Adrar, Algeria. The proposed system is based on adaptive combination of two different models. The first model is an ARMA-based model, while the second model is an artificial neural network ANN-based model. To allow adaptive combination, models are associated to time varying coefficients that are updated recursively using the recursive least square algorithm (RLS). Numerical simulations have shown that for few hours in advance, the prediction error of the combined system is lower or at least equal to the best estimator.

Keywords: autoregressive moving average process; artificial neural networks; recursive least square method; wind speed forecasting; combined models

Citation: Chellali, F. Short-Term Wind Forecasting in Adrar, Algeria, Using a Combined System. *Eng. Proc.* **2023**, *29*, 11. https://doi.org/10.3390/engproc2023029011

Academic Editors: Abdelmadjid Recioui, Hamid Bentarzi and Fatma Zohra Dekhandji

Published: 13 January 2023

1. Introduction

In wind energy industry, there is always some uncertainty about the final product due to the fact that wind speed is highly variable. The ability to predict wind speed for few hours in advance will help to ensure efficient utilization of the power generated, and therefore, enhance the position of wind energy compared to other forms of energy.

Forecasting of wind speed has been the subject of a lot of studies. Statistical and artificial neural networks are the approaches the most found in literature. Based on the hourly averaged wind speed (HAWS), several models have been developed using time series methodologies. In 1984, Geerts has proposed a short-term forecasting of wind speed using ARMA models [1]. A comparison of the performances of ARMA models developed using one-year data with those of the persistent model has indicated that for more than 1 h, ARMA models provide better forecasting. This shows that models developed using several years data could provide better results than those developed using only one-year data. Daniel and Chen (1991) have used a three-year long time series to develop ARMA models [2]. Since then, several similar studies have been realized for many sites around the world. Using 12-year data, Nfaoui et al. (1996) concluded that an AR(2) model is able to simulate the wind speed data of Tangiers (Morocco) [3]. Such methodologies have been confirmed by Kamal and Jafri (1997) using data of Quetta (Pakistan) [4]. Torres et al. (2004) used data of five locations in Navarre (Spain) to identify up to 10 different ARMA models [5].

Models based on artificial neural networks have been compared with ARMA models by Sfetsos (1999) [6]. Using only data of one month (March), the author has concluded that ANN models outperform the linear models. Using ANN and ARIMA models, More et al. (2003) have forecasted daily and monthly wind speed in India. Results indicate that performances of ANN models are better than ARIMA models [7]. Using one-month long data (January), Erasmo et al. (2009) concluded that for the particular case of La Venta (México), a model with two layers and three neurons was the best for training and forecasting [8].

In the following work, we propose HAWS forecasting systems based on adaptive combinations of two alternative (individual) models. The first model is based on ARMA approach, while the second is based on ANN approach. Adaptive combination is done by associating time-varying weights to the alternative models. The time-varying weights are adapted recursively using recursive least square algorithm (RLS). An important motive to combine forecasts from different models is the fundamental assumption that one model cannot identify the true process exactly, but different models may play a complementary role in the approximation of the data generating process, especially in case of doubt about the existence of the best estimator.

Data used in this work are measured over four years (November 2004 to October 2008) by the meteorological station situated at the airport of Adrar, Algeria (27.9° N, 0.3° W). Because of the gaps found in time series, our study is limited only for three months (January, November, and December). Data of the three first years are used to identify ARMA and train ANN models. The four-year data are used as an independent dataset to verify the forecasting ability of the obtained models.

2. ARMA Modeling

ARMA modeling for HAWS consists of three main steps: the first step is the power transformation which is done in order to carry the wind speed from Weibull distribution to Gaussian distribution [2]. The second step is the standardization step. The purpose of this step is to eliminate non-stationarity due to daily cyclical behaviors. The last step is the identification step which consists of the order determination and parameters estimation.

2.1. Power Transformation

For lot of sites over the word, it has been found that Weibull distribution (Equation (1)) fits the wind distribution the best:

$$f(v) = \frac{k}{c}\left(\frac{v}{c}\right)^{k-1} e^{-\left(\frac{v}{c}\right)^k} \tag{1}$$

where v is the wind speed, k is the form factor, and c is the scale factor.

In order to fit ARMA models to the HAWS, power transformation of the observed data must be performed. The aim of the power transformation is to approximate the distribution of the wind data from a Weibull distribution to Gaussian one. The power transformation is performed by raising each value of the observed data by the same power index. For more accurate approximation, Daniel and Chen (1991) have proposed to use $x = k/3.6$ as a reference to obtain more accurate index m using skewness statistics is given as [2]:

$$S_m = \sum_{y=1}^{Y}\sum_{n=1}^{N} \frac{\left[\left(v_{n,y}^m - mean(v_{n,y}^m)\right)/std(v)\right]^3}{Y.N} \tag{2}$$

where Y is the number of considered years and N is the number of samples per month. After iterative calculations using several values of m, the selected m is the one that makes distribution of v^m symmetric i.e., $S_m \approx 0$. Values of Weibull parameters, of Dubey's index x and asymmetry index m are evaluated for the three proposed months and presented in Table 1.

Table 1. Estimated parameters for the Weibull distribution and ARMA models.

Title 1	Month	C	K	x	m	ARMA	AR	MA
January	6.43	7.19	2.55	0.70	0.68	(3,0)	$ARMA(3,0)$ $\phi_1 = 0.70$ $\phi_2 = 0.10$ $\phi_3 = 0.06$	
November	5.89	6.61	2.23	0.62	0.75	(3,0)	$\phi_1 = 0.72$ $\phi_2 = 0.09$ $\phi_3 = 0.07$	-
December	6.02	6.74	2.22	0.63	0.57	(3,0)	$\phi_1 = 0.72$ $\phi_2 = 0.11$ $\phi_3 = 0.04$	-

2.2. Standarisation

Diurnal non-stationarity can be eliminated by subtraction the hourly averaged wind speed $u(t)$ from the transformed HAWS then dividing by the hourly standard deviation $s(t)$ [5]. Transformed and standardized HAWS (TS-HAWS) are given as:

$$v^*_{n,y} = (v^m_{n,y} - u(t))/s(t) \tag{3}$$

where

$$u(t) = \frac{\sum\limits_{y=1}^{Y} \sum\limits_{i=1}^{d} v^m_{y,i*24+t}}{Y * d} \tag{4}$$

and

$$s(t) = \sqrt{\frac{\sum\limits_{y=1}^{Y} \sum\limits_{i=1}^{d} \left(v^m_{y,i*24+t} - u(t)\right)^2}{Y * d}} \tag{5}$$

It is assumed that $u(t)$ and $s(t)$ are periodic functions, i.e., $u(1) = u(25)$, $u(2) = u(26)$, $s(1) = s(25)$, and $s(2) = s(26)$, where d is the number days for given month.

To illustrate the effect of standardization on HAWS, periodograms of original HAWS and TS-HASW are evaluated and presented in Figure 1. It is clear that diurnal and semi-diurnal harmonics present in the periodogram of HAWS in the form of peaks have been canceled from the periodogram of the TS-HAW.

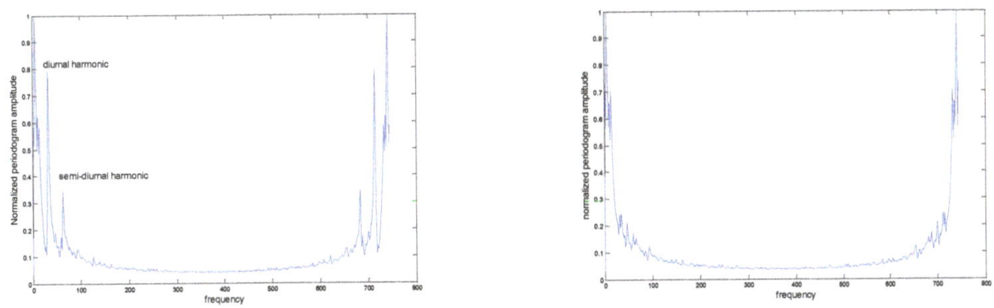

(a) (b)

Figure 1. Periodograms of HAWS (**a**) and TS-HAWS (**b**).

2.3. ARMA Fitting

The following section consists of fitting ARMA models to TS-HAWS. A general model is given as:

$$\begin{aligned} v^*_{n,y} &= (\phi_1 L + \phi_2 L^2 + \cdots + \phi_p L^p) v^*_{n,y} \\ &+ \varepsilon_{n,y}(1 - \theta_1 L - \theta_2 L^2 - \cdots - \theta_q L^q) \end{aligned} \tag{6}$$

where p is the order of the autoregressive part, q is the order of the moving part, L is the lag operator, and ε is a white Gaussian noise of zero mean and variance σ^2_ε.

2.3.1. Order Determination

The following section consist of identifying the values of p and q. A pure MA process exhibits a cut of after q lags in the autocorrelation function (ACF); however, for pure AV or mixed ARMA process, the ACF deceases exponentially. Order of AV part can be determined using the partial autocorrelation function (PACF) that cut off after p lags for a pure AV process while it dies gradually in case of pure MA process.

Figures 2 and 3 present the ACF and PACF, respectively, evaluated for January, November, and December. While the ACFs of the three proposed months decrease exponentially, the PACFs cut almost zeros after the third order; this implies that HAWS can be modeled by a low order $ARMA(p, q)$.

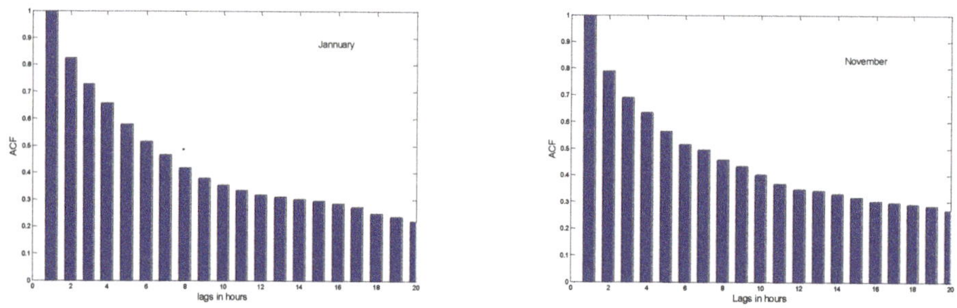

(a) **(b)**

Figure 2. Autocorrelation function of HAWS for (**a**) January and (**b**) November.

(a) **(b)**

Figure 3. Partial Autocorrelation function of HAWS for (**a**) January and (**b**) November.

ACF and PACF are used to determine an adequate group of ARMA models. Appropriate orders (p, q) are determined with the help of an additional criterion, such as Akaike information criterion [Storres].

$$AIC(p, q) = (Y.N) \cdot \ln\left(\sigma^2_\varepsilon(p, q)\right) + 2T) \tag{7}$$

where T is the total number of parameters to be estimated.

2.3.2. Parameters Estimation Phase

Once p and q have been identified. The model coefficients ϕ_i and θ_i along the variance of the residuals σ_ε^2 can be estimated. The preliminary estimation is done applying the Yule–Walker relations for ϕ_i, while θ_i values are obtained using Newton–Raphson Algorithm. Final estimation of the parameters using the method of least squares error. Selected models for each month are presented in Table 1.

Note that for the three studied months, it has been found that $ARMA(3,0)$ is the best model (Table 1).

Finally, the models are validated by evaluating the ACF of the residuals. The fitted model can be accepted if the residuals are uncorrelated and normally distributed (Figure 4).

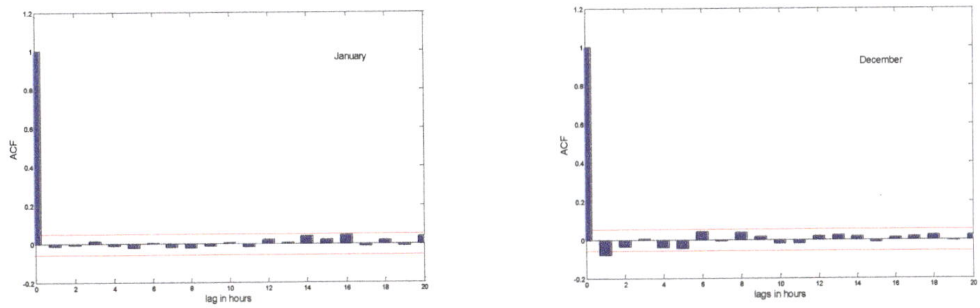

(a) (b)

Figure 4. Autocorrelation function of the residues HAWS for (**a**) January and (**b**) November.

In order to validate the estimated models, we evaluated the residual autocorrelation functions (Figure 5). If the residuals are jointly independent, their autocorrelation functions cancel for a lag τ = 0. From Figure 5, we can see that the residuals are uncorrelated. Thus the models are retained.

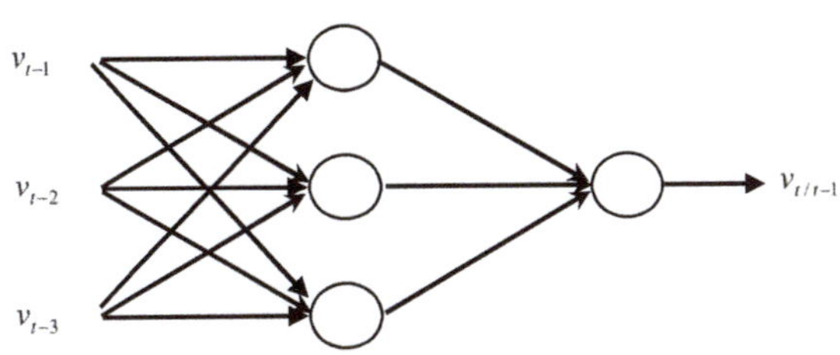

Figure 5. Neural network architecture with four neurons and two layers.

3. The Persistent Model

Persistent model as defined in [5] is given as:

$$v^P{}_{t+h/t} = v_t \tag{8}$$

Equation (8) is equivalent to saying that the wind at instant $h + t$ is simply the same as it was at time t. This model is developed by meteorologists as a comparison tool to supplement the other models. The accuracy of this model decreases rapidly with an increase of prediction lead time.

4. Neural Networks

Thus far, we have examined linear prediction with ARMA models, which constitute a mostly linear approach to data analysis. Now, we turn to the more complex neural networks. Neural networks have similar uses to linear prediction filters, and can be applied to the same general set of problems.

In general, rules to determine the number of inputs, outputs, and layers do not exist. In the following work, and based on results obtained by Sfetsos (1999) and Erasmo (2009) [9,10], it was decided to use feed-forward ANN (Figure 6). In fact, Sfetsos has found that with the Levenberg–Marquard algorithm as a training method, the feed-forward ANN is very adequate to model the wind speed.

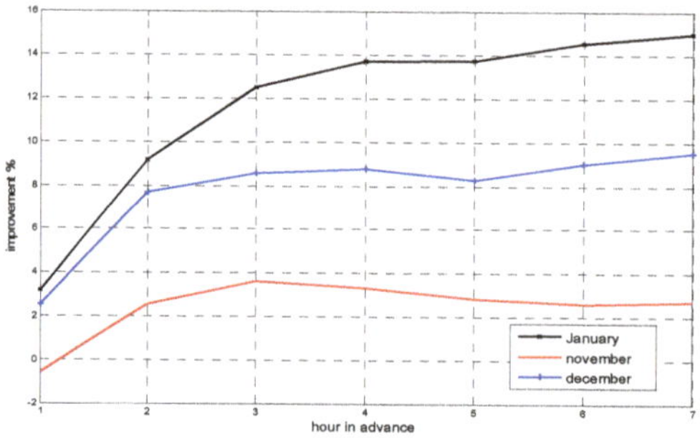

Figure 6. Percentage improvement of RMSE of the ANN forecasts when TS-HAWS are used opposed to HAWS.

The response of the ANN output neuron is:

$$v_t = f_2\left(\sum_{q=1}^{3} w_q f_1\left(\sum_{j=1}^{3} w_{q,j} v_{t-j} + bj\right) + b_q\right) \tag{9}$$

where f_1 and f_2 are nonlinear sigmoid activation functions.

The proposed ANN networks are trained to predict the wind speed one step ahead. Prediction of h hours in advance is obtained iteratively, i.e., $v_{t+2/t}^{(ANN)}$ is predicted using $v_{t+1/t}^{(ANN)}$, $v_{t+3/t}^{(ANN)}$ is predicted using $v_{t+3/t}^{(ANN)}$, and so on.

5. Forecast Combination

When several candidate models are available to forecast single variable, we can either select the best model or combine them. Combination of forecast is very advised in case of doubt about the existence of the best model. Our purpose in the following section is to build a linear combination $v_{t/t-h}^{Comb}$ of the competing forecasts as:

$$v_{t/t-h}^{Comb} = w_t^{(ARMA)} v_{t/t-h}^{(ARMA)} + w_t^{(ANN)} v_{t/t-h}^{(ANN)} \tag{10}$$

where $w_t^{(ARMA)}$ and $w_t^{(ANN)}$ are time varying coefficients that are updated recursively using recursive least square method (RLS). RLS estimation consists of minimizing the cost function given as:

$$\begin{aligned} S_t^2(\mathbf{w}) &= \sum_j^t \kappa(t,j)\left(v_{j/j-h}^{Comb} - \mathbf{w}\cdot\mathbf{v}_{j/j-h}\right)^2 \\ &= S_{t-1}^2(\mathbf{w}) + \kappa(t,t)\left(v_{t/t-h}^{Comb} - \mathbf{w}\cdot\mathbf{v}_{t/t-h}\right)^2 \end{aligned} \tag{11}$$

where $\mathbf{w} = \left[w^{(ARMA)} \quad w^{(ANN)} \right]$, $\mathbf{v}_{t/t-h} = \left[v_{t/t-h}^{(ARMA)} \quad v_{t/t-h}^{(ANN)} \right]$ and $\kappa(t,j)$ is the forgetting profile. Usually, $\kappa(t,j) = \Pi_{i=j+1}^{t} \lambda_i$, where $0 \leq \lambda_i \leq 1$ is the forgetting factor [11] The forgetting profile $\kappa(t,j)$ is the weight associated to the jth residual and it allows to reduce the importance of old data recursively.

According to [11], the RLS estimator of $\mathbf{w_t}$ is given as:

$$\mathbf{w_t} = \mathbf{w_{t-1}} + \mathbf{\Gamma_t} \mathbf{v_{t/t-1}} a_t \tag{12}$$

where $a_t = v_t - \mathbf{v_{t/t-1}} \mathbf{w_{t-1}}$, which is the one step-ahead prediction error, and $\mathbf{\Gamma_t}$ is called the gain matrix or the weighted covariance matrix that can be estimated as [12]:

$$\mathbf{\Gamma}_t = \frac{1}{\lambda} \left(\mathbf{\Gamma}_{t-1} - \frac{\mathbf{\Gamma}_{t-1} \mathbf{v_{t/t-1}} \mathbf{v_{t/t-1}}' \mathbf{\Gamma}_{t-1}}{\lambda + \mathbf{v_{t/t-1}}' \mathbf{\Gamma}_{t-1} \mathbf{v_{t/t-1}}} \right) \tag{13}$$

The choice of λ is very important in the recursive estimation of weights. It is indicated in [13] that most applications use a constant forgetting factor typically inside $0.950 \leq \lambda \leq 0.999$.

6. Results and Discussion

To allow a comparison of models, the evolution of the RMSE is evaluated for the persistent, ARMA, feed-forward, and combined models when forecasts are done 1–7 h in advance.

For ANN models, it has been found that forecasting wind speed using TS-HAWS instead of HAWS data yields better performances. As indicated in Table 2, only in one case, forecasting with HAWS has provided lower RMSE (1 h ahead in November). Improvement of forecasting with TS-HAWS over HAWS is presented in Figure 6. Maximum improvement of 15% has been obtained in January when forecasting 7 h ahead.

Table 2. Prediction performances comparison between transformed and non-transformed data.

Hours in Advance	RMSE (January)		RMSE (November)		RMSE (December)	
	HAWS	TS-HAWS	HAWS	TS-HAWS	HAWS	TS-HAWS
1	1.2933	1.2522	1.3188	1.3264	1.4032	1.3675
2	1.6143	1.4666	1.6987	1.6551	1.8266	1.6864
3	1.8770	1.6426	1.9715	1.9008	2.1000	1.9197
4	2.0053	1.7306	2.1686	2.0973	2.2858	2.0847
5	2.1302	1.8374	2.2954	2.2302	2.4260	2.2246
6	2.2145	1.8922	2.3655	2.3038	2.5648	2.3334
7	2.2723	1.9322	2.4007	2.3359	2.6599	

For all the studied cases, we have found that the RMSE of the persistence models are greater than the RMSE of the other models (Table 3). These results are similar to those obtained by Torres [5] and Sfetsos [6], which indicated that ARMA- and ANN-based models provide better forecasts than the persistence models.

Table 3. Prediction performances for January.

Hour in Advance	RMSE in m/s				Improvement over the Persistent Model %		
	Persistent	ARMA	ANN	Combined	ARMA	ANN	Combined
1	1.3676	1.2352	1.2522	1.2347	9.6845	8.4387	9.7188
2	1.7368	1.4546	1.4666	1.4542	16.2518	15.5609	16.2737
3	2.0730	1.6315	1.6426	1.6311	21.2987	20.7610	21.3178
4	2.2489	1.7308	1.7306	1.7310	23.0383	23.0483	23.0303
5	2.4499	1.8341	1.8374	1.8343	25.1331	25.0000	25.1272
6	2.5899	1.8922	1.8922	1.8926	26.9397	26.9372	26.9237
7	2.7073	1.9361	1.9322	1.9368	28.4834	28.6281	28.4596

In 66.67% of the studied cases, it has been found that ARMA models provide relatively lower RMSE than ANN models (Tables 3 and 4). Evaluation of the forecasting improvement of ARMA models over ANN models varies between −0.61% and 03.49%. These results differ from that obtained by Strores (1999), who has found that ANN models can achieve lower RMSE than ARMA models.

Table 4. Prediction performances for November.

Hour in Advance	RMSE in m/s				Improvement over the Persistent Model (%)		
	Persistent	ARMA	ANN	Combined	ARMA	ANN	Combined
1	1.3621	1.3114	1.3264	1.3104	3.7196	2.6187	3.7928
2	1.7986	1.6550	1.6551	1.6494	7.9842	7.9780	8.2972
3	2.1484	1.8999	1.9008	1.8949	11.5639	11.5260	11.7971
4	2.4454	2.0983	2.0973	2.0912	14.1958	14.2358	14.4833
5	2.6646	2.2245	2.2302	2.2184	16.5186	16.3015	16.7471
6	2.8103	2.2948	2.3038	2.2888	18.3407	18.0203	18.5543
7	2.8887	2.3252	2.3359	2.3179	19.5068	19.1383	19.7606

Combined models provide relatively better performances than the ARMA and the ANN models a few hours in advance (1–4 h); however, for long-term forecasting, individual models can provide lower RMSE than the combined models (Table 3).

To better understand the role of individual models in the final combination, we have presented in Figure 7 the variation of the combination coefficients when real wind is measured for November and December 2007. The first 100 wind speed values have been used to initialize coefficients and the covariance matrix. Figure 7a shows that the coefficient of the ARMA model is greater than the one of the ANN model, which means that the ARMA model is providing lower error for November 2007. In December 2007, and as indicated in Figure 7b, the ANN model forecasts the wind speed better than ARMA models.

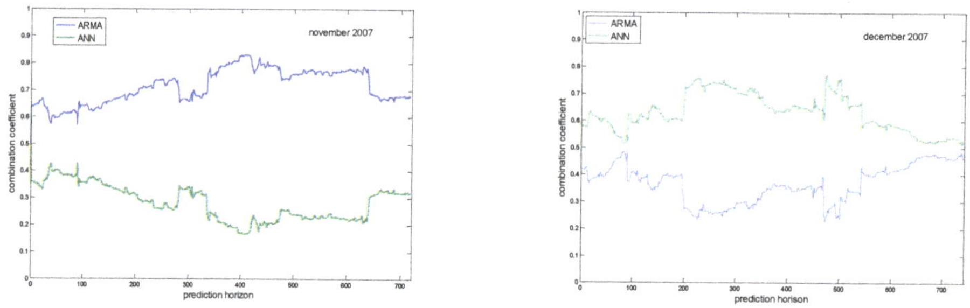

(a) (b)

Figure 7. Evolution of the combination coefficient for November and December.

7. Conclusions

In this study, models based on ARMA methodology as well as ANN theory have been proposed to forecast the hourly wind speed. The obtained results have indicated that both ARMA and ANN models are capable to predict wind speed for few hours in advance. Performances comparison between ARMA and ANN models has indicated that their performances are very close.

An adaptive combination of the two proposed approaches has proposed to improve the forecasting performances. The obtained results have indicated that for a few hours in advance, the combined model outperform the individual models.

For ANN modeling, it has been found in this work that the use of the transformed and standardized TS-HAWS provides better results than HAWS.

Funding: This research received no external funding.

Institutional Review Board Statement: Not applicable.

Informed Consent Statement: Not applicable.

Data Availability Statement: Data sharing not applicable. No new data were created or analyzed in this study. Data sharing is not applicable to this article.

Conflicts of Interest: The authors declare no conflict of interest.

References

1. Geerts, H.M. Short range prediction of wind speeds: A system-theoretic approach. In Proceedings of the European Wind Energy Conference, Hamburg, Germany, 22–26 October 1984; pp. 594–599.
2. Daniel, A.; Chen, A. Stochastic simulation and forecasting of hourly average wind speed sequence in Jamaica. *Sol. Energy* **1991**, *46*, 1–11. [CrossRef]
3. Nfaoui, H.; Essiarab, H.; Sayigh, A.H. A stochastic Markov chain model for simulating wind speed time series at Tangiers, Morocco. *Renew. Energy* **2004**, *29*, 1407–1418. [CrossRef]
4. Kamal, L.; Jafri, Y.Z. Time series models to simulation and forecast hourly averaging wind speed in Quetta, Pakistan. *Sol. Energy* **1997**, *61*, 23–32. [CrossRef]
5. Torres, J.; Garcia, A.; De Blas, M.; De Francisco, A. Forecast of hourly averaged wind speed with ARMA models in Navarre (Spain). *Sol. Energy* **2005**, *79*, 65–77. [CrossRef]
6. Sfetsos, A. A comparision of various forecasting techniques applied to mean hourly wind speed time series. *Renew. Energy* **2000**, *21*, 23–35. [CrossRef]
7. More, A.; Deo, M. Forecasting wind with neural networks. *Mar. Struct.* **2003**, *16*, 35–49. [CrossRef]
8. Erasmo, C.; Wilfrido, R. Wind speed forecasting in the South Coast of Oaxaca, México. *Renew. Energy* **2007**, *32*, 2116–2128.
9. Sfetsos, A. A novel approach for the forecasting of mean hourly wind speed time series. *Renew. Energy* **2002**, *27*, 163–174. [CrossRef]
10. Erasmo, C.; Wilfrido, R. Short term wind speed forecasting in la Venta, Oaxaca, México, using artificial networks. *Renew. Energy* **2009**, *34*, 274–278.
11. Sanchez, I. Adaptive combination of forecasts with application to wind energy. *Int. J. Forecast.* **2008**, *24*, 679–693. [CrossRef]
12. Grillenzoni, C. Optimal recursive estimation of dynamic models. *J. Am. Stat. Assoc.* **1994**, *89*, 777–787. [CrossRef]
13. Sanchez, I. Short-term prediction of wind energy production. *Int. J. Forecast.* **2006**, *22*, 43–65. [CrossRef]

engineering proceedings

MDPI

Proceeding Paper

Implementation of Battery Characterization System †

Abdelaziz Zermout, Hadjira Belaidi * and Ahmed Maache

Signals and Systems Laboratory, Institute of Electrical and Electronic Engineering, University M'hamed Bougara of Boumerdes, Boumerdes 35000, Algeria
* Correspondence: ha.belaidi@univ-boumerdes.dz or hadjira983@yahoo.fr

† Presented at the 2nd International Conference on Computational Engineering and Intelligent Systems, Online, 18–20 November 2022.

Abstract: The successful transfer toward green renewable energy depends heavily on good, reliable Energy Storage Systems (ESS). Lithium-ion batteries are the preferred choice for many applications; however, they need careful management, especially an accurate State-Of-Charge (SOC) estimation. Hence, in this paper an overview of some SOC estimation methods is briefly described; then, an automated battery cell test system prototype that will enable further improvement is designed and implemented. Some tests are conducted on an aged lithium-ion cell and the obtained results are satisfactory and accurate with an error of around 0.5×10^{-3} (V or A), thus validating the proposed prototype.

Keywords: Energy Storage Systems (ESS); State-Of-Charge (SOC); state-of-health; lithium-ion battery; Battery Management System (BMS); battery test; battery characterization prototype

Citation: Zermout, A.; Belaidi, H.; Maache, A. Implementation of Battery Characterization System. *Eng. Proc.* **2023**, *29*, 12. https://doi.org/10.3390/engproc2023029012

Academic Editors: Abdelmadjid Recioui, Hamid Bentarzi and Fatma Zohra Dekhandji

Published: 16 January 2023

1. Introduction

Lithium-ion cells can be considered the backbone of many Battery Energy Storage Systems (BESS). They have many relative advantages when compared to other cells, such as power density, power throughput, high lifecycle, and low self-discharge; however, they need careful management. Lithium-ion cells are available in numerous types and brands with different characteristics. By connecting them in series and/or parallel, these cells are used to construct batteries of different voltages and capacities, from a few watt hours to hundreds of Megawatt-hours. These are widely used in various applications, such as electric vehicles (EVs), storage of renewable energy, power banks, electric yachts, bikes, scooters, laptops, smartphones, and many other applications. In each application, lithium-ion batteries are used in different ways and in different conditions. In order to manage and run them effectively and securely, many parameters should be taken into consideration. One of the important parameters that should be known and monitored in real time is the state of charge (SOC), which represents the amount of capacity left in the battery (in Ah) as a percentage of its maximum capacity. In addition, it is important to estimate the state of health (SOH) of a battery, which represents the total amount of energy that the battery can hold and deliver compared to a fresh battery [1–3].

Because of the global transfer toward renewable green energy, and increasing demand of BESS-based systems like EVs and smart grids, many developments have been undertaken to BMS (Battery Management System) technology to become more efficient and reliable. Moreover, to provide good management of the batteries, particularly SOC and SOH estimations, many techniques have been proposed by researchers and developers for SOC estimation. Some of the approaches are enhancing the existing techniques and some of them are more recent, such as Data-Driven Model-based methods using Machine Learning (ML) that show very interesting results [1,4,5]. However, although the noticed improvements in SOC estimation methods are important, there is as yet no universal solution for SOC estimation sufficiently accurate for every battery in any condition; therefore, more research and development are still to be done in this field.

Wrong estimation of lithium-ion batteries' SOC can lead to sudden failure or complete shutdown of any BESS-based system, and rapid degradation of the batteries' ability to store and deliver energy due to overcharge or discharge that can occur, potentially causing an unsafe situation. The estimation of SOC is not a direct measurement, but is a tricky operation and depends on many other parameters like voltage, current, temperature, and aging. There are various methods used to estimate the SOC. The criteria to choose the best method for a certain application is that it should be an online method without affecting the performance of the batteries, and with less computational complexity and high estimating accuracy [2,6]. Hence, this paper presents a prototype battery characterization system that will be used as a low-cost laboratory test bed in order to develop and test advanced estimation techniques for SOC and SOH estimations.

2. State of Charge Estimating Methods Overview

Due to the recent increased need for BESS-dependent systems such as EVs and renewable ESS, several research works and ongoing development have been undertaken to improve the performance and the reliability of such systems, by making battery management systems more sophisticated and more accurate. One of the main tasks of the battery management system is to accurately control SOC and SOH as much as possible in real time to prevent any unwanted scenarios. Thus, many methods have been developed to estimate the SOC, and each of them can be more suitable than the others for certain applications.

2.1. Open Circuit Voltage (OCV)

Open circuit voltage is getting the SOC of the battery by converting its open circuit voltage to its state of charge using a predefined look-up table that maps the OCV to SOC. This method is quite precise; however, it is not too practical in real applications. In fact, the battery should be at rest for a large amount of time until the stabilization of its electrochemistry; then, the OCV measurement can be performed and compared to the look-up table. The drawback of this method is that one look-up table cannot be used for all batteries; mostly, it is affected by the battery itself, its type, and its aging [7,8].

2.2. Coulomb Counting

This method is very common and simple. It depends on counting the amount of the charge flow in or out of the battery in (*Ah*) by integrating the current in time. This method can be expressed by Equation (1) where I is the current entering the battery, SOC (t = 0) is the initial estimated SOC, and C max is the maximum capacity that the battery is able to store and deliver.

$$SOC(t) = SOC(t = 0) + \frac{1}{C\,max} \int_{t=0}^{t} Idt \tag{1}$$

An accurate estimation of this method depends directly on high-precision measurement of the current and knowing the exact initial state of charge SOC (t = 0) and the state of health SOH. Therefore, any slight errors that will occur even with high-precision measurements will cause an accumulated error over time; in the process, it will cause a large deviation in the SOC estimation [2,3,9].

2.3. Voltage Method

This depends on the different existing discharging curves that are generated experimentally for certain batteries. It is not computationally complex and suitable for constant current applications and constant conditions in general; however, when the current or temperature is fluctuating, its accuracy will decline and it will no longer be practical. Thus, it is not very practical for real applications [1,2].

2.4. Kalman Filter

Kalman filter is an algorithm that can estimate the inner state of many systems; as well as this, it can be used for SOC estimation of a battery with a suitable model. It is based on the current, voltage, and temperature measurements to estimate the state of charge [1]. EKF (extended Kalman Filter) compares the cell voltage that was actually measured to the cell voltage that a battery model predicted to estimate the battery SOC after corrections [10].

2.5. Equivalent Circuit Model (ECM)

This technique presents the battery as an electric equivalent circuit model by choosing the right circuit and the right values of its components, so that it can predict the parameters of the battery in real time. Some circuit models are shown in Figures 1 and 2. This method has acceptable results, it is simple and computationally less complex; however, after the battery aging, the model chosen can no longer fit the battery behavior [1,11].

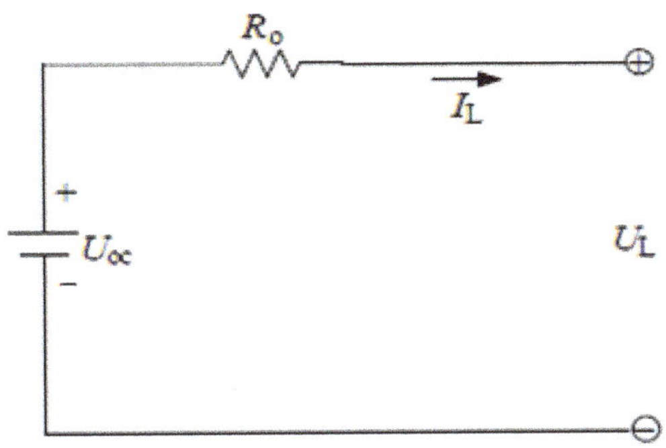

Figure 1. Rint model circuit diagram [11].

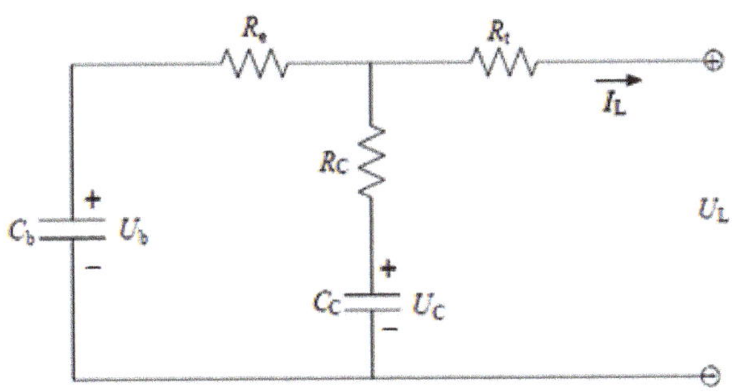

Figure 2. RC model circuit diagram [11].

2.6. Impedance Spectroscopy

The internal impedance spectroscopy of lithium-ion batteries can reflect the SOC with a high accuracy. By analyzing the impedance of the battery at different frequencies from 40 Hz to 20 kHz, the phase and the modulus is measured and compared to predefined impedance spectroscopy to get the SOC. Conversely, it is implemented using expansive hardware; as well as interrupting the battery performance, the external condition and aging my affect its precision [1,12].

2.7. Data-Driven Model

This method does not depend on any physical model but relies on using artificial intelligence (AI), where a suitable neural network can be implemented and trained with sufficient related data of the battery behavior. This technique gained the advantage with the recent improvement in computational power of the hardware and the massive data that is generated from different applications like EVs and other BESS systems. However, this method is computationally complex and has an over-fitting problem [1,4,13].

3. Problematic

Our state-of-the-art overview of the existing SOC estimation methods led us to summarize the advantages and drawbacks of each technique in Table 1.

Table 1. Advantages/drawbacks of SOC estimation methods.

Method	Advantages	Disadvantages
Open circuit voltage	Simple and accurate	The battery should be at rest for a long time
Coulomb counting	Common and simple	It will have an accumulated error over time
Voltage method	Easy and simple	It is affected by fluctuating currents
Equivalent circuit model	The model is easy to construct	It may not reflect the behavior of the battery over time
Kalman filter	Accuracy is good	The accuracy depends on the used model
Impedance spectroscopy	Accurate	Need expansive equipment and its accuracy is affected by temperature and aging
Data-driven method	Accurate and no model is needed	High compute complexity and needs a lot of related data

Therefore, due to the diversity of the types and brands of Lithium-ion cells that are existing today and the different conditions and applications where they can be used; there is a need for further investigation and improvement of SOC estimating methods. Many tests should be conducted in regulated conditions, and a lot of data should be generated. Thus, an affordable Battery cell test system is needed. This article deals with the designing, implementation, and demonstration of a low-cost prototype battery cell test system that can perform several types of testing with pretty accurate results. At the same time, it offers good flexibility and ergonomic use; thus, this will provide more accessibility for more experiments and tests to be done.

4. Design and Implementation

Lithium-ion cell is not a simple component that has characteristics that can be known or predicted easily; as well, these characteristics are not linearly correlated. This is due to its complex electrochemistry behavior in different conditions, and because of the different types of batteries that already exist and keep emerging every day. Consequently, many tests should be conducted in a controlled environment to classify and understand more about the characteristics of the batteries. These tests can include capacity tests, lifecycle tests, aging, internal resistance, the best temperature for different operations, optimum charging and discharging current, and testing of the maximum safe limits of the battery operation in controlled conditions to ensure safe operation in the real world. Thus, these tests are essential for battery manufacturers, developers, and researchers. The battery can be in different states: charging, discharging, or at rest (no current flow). At the same time, the temperature can be varying, and other parameters can be included too. In order to test the battery in the mentioned states, the battery tester should be automated and consist of at least: a control unit, data acquisition unit, controlled variable load unit, controlled charging unit, and possibly adding thermal management unit [14,15] as demonstrated in Figure 3.

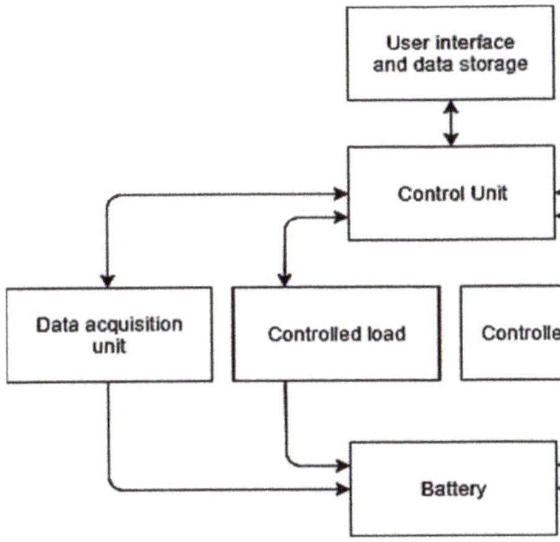

Figure 3. Overall diagram of a battery test system.

The designed system block diagram is shown in Figure 4. It consists of different components. ESP32 is the main control unit that controls the system. It has an 80 MHz processor, and 38 IO pins. It communicates with the ADS1115 and controls the LCD screen, the relays, discharging MOSFET, and charging PNP Transistor using PWM. ADS1115 is a 16-bit ADC that converts the analog signal to 16-bit 2′s complement format. It senses voltage 1, which is the voltage of the battery, and the difference between voltage 1 and voltage 2 that will be divided by the shunt resistor to get the current. The LCD display and push buttons will be the main user interface, especially when a computer is not connected. The MicroSD card will be used for data storage, as well as the test instructions that can be saved and executed without the need for a computer. For the temperature-controlled chamber, it has not yet been completely implemented. It will consist of two temperature sensors–one for the chamber, and one for the battery surface. The implemented system is shown in Figure 5; the components used are outlined in Table 2.

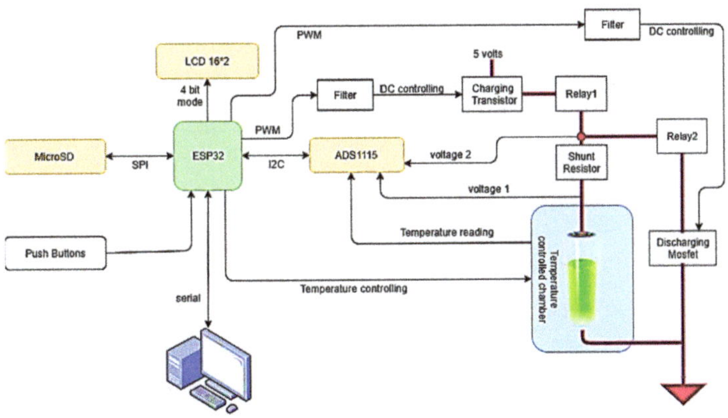

Figure 4. The diagram of the designed battery test system. The LCD is 16 * 2 i.e., 16 columns by 2 rows so it can display 16 × 2 = 32 characters in total.

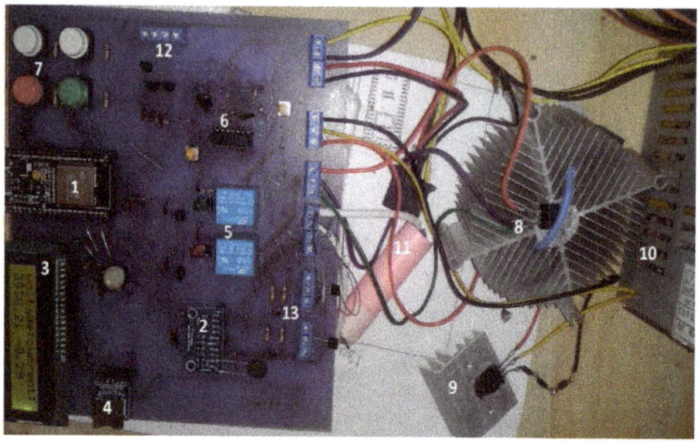

Figure 5. The implemented hardware. The numbers refer to the different used components which are outlined in Table 2.

Table 2. The used component referred in Figure 5.

N°	Description
1	ESP32
2	ADS1115
3	16*2 LCD display
4	MicroSD card
5	Relays 1 and 2
6	Low-pass RC filters
7	Push buttons
8	Discharging MOSFET
9	Charging transistor
10	Power supply
11	Lithium-ion cell
12	Connections for controlling the temperature chamber
13	Temperature sensors for the temperature-controlled chamber

5. Results and Discussion

A number of tests were performed on an aged lithium-ion cell at room temperature (around 24 °C) with the characteristics described in Table 3.

Table 3. Characteristics of the lithium-ion battery used in our tests.

Manufacturer	Model	Rated Capacity	Maximum Voltage	Minimum Voltage
LG	LGABD18650	3000 mAh	4.35 V	3.0 V

A Multistage Constant Current (MSCC) charging protocol was applied to the lithium-ion cell. The results are shown in Figure 6, which resemble the results obtained by the study in [16] and are illustrated in Figure 7. During this test, the battery was charged with a constant current of 1 A until it reached 4.35 V; then, the current was stepped down to a less constant current. Thus, the voltage of the battery drops down; it is kept until it reaches 4.35 V again. The same steps were repeated until the current is less than or equal to 100 mA; then, the charging process was stopped.

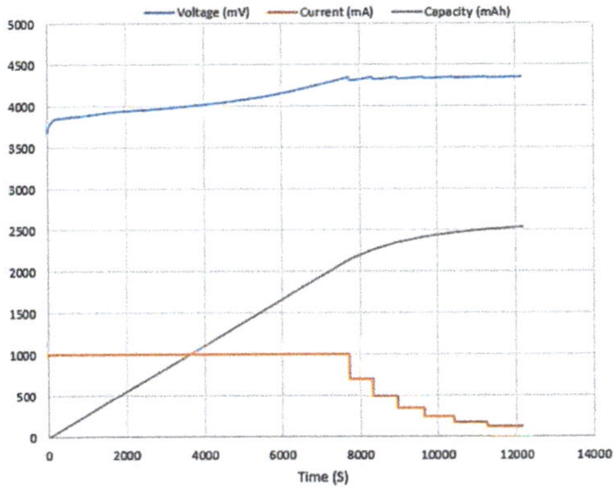

Figure 6. Charging with MSCC protocol.

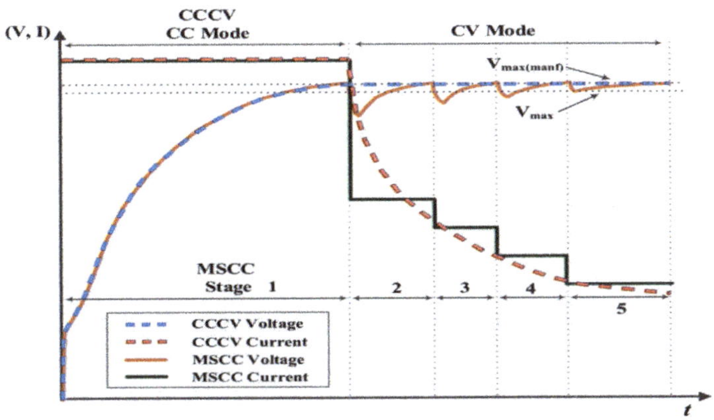

Figure 7. MSCC charging protocol [16].

Figure 8 shows the discharging curve. The battery is discharged with 1 A constant current until the voltage reached the cutoff voltage of 3.0 V. The x-axis represents the total capacity discharged in (mAh). Figures 9 and 10 represent the voltage response of the battery to discharging pulse and charging pulse respectively, with the negative current denoting the discharging current.

Table 4 shows the performance characteristics of the implemented hardware. As illustrated in the results, the voltage and current resolution is less than 2 m (V or A) with a sampling rate of 35 samples/s. This led to satisfactory outcomes with a very small reading error of approximately 0.5 m (A or V). Thus, the implemented hardware can be used for some battery characterization purpose.

Table 4. Performance characteristics of the implemented hardware.

Voltage Resolution	Current Resolution	Maximum Sampling Rate	Maximum Voltage/Current	Load Hysteresis at 1 A	Charging Hysteresis at 1 A	Voltage Error	Current Error
0.187 mV	0.153 mA	35 Samples/s	5 V/5 A	<3.5 mA	<3 mA	0.6 mV	0.5 mA

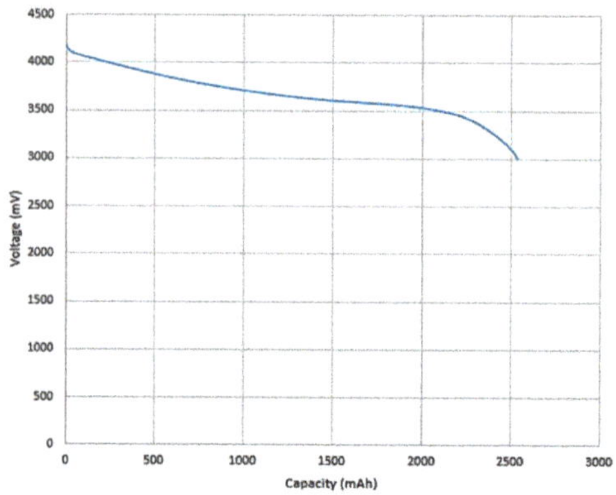

Figure 8. Discharging curve with 1 A at 24 °C.

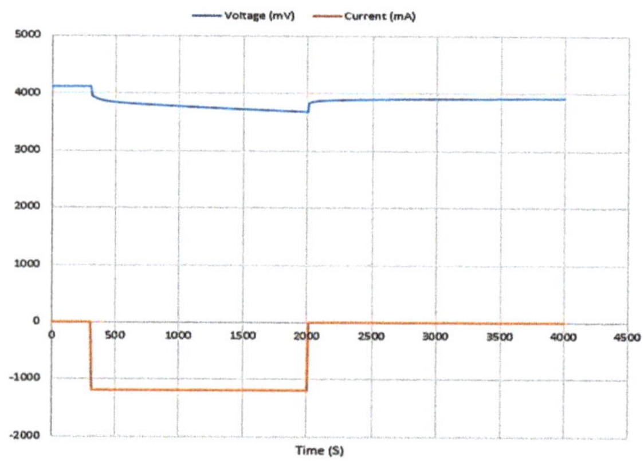

Figure 9. Voltage response of the battery to discharging current pulse.

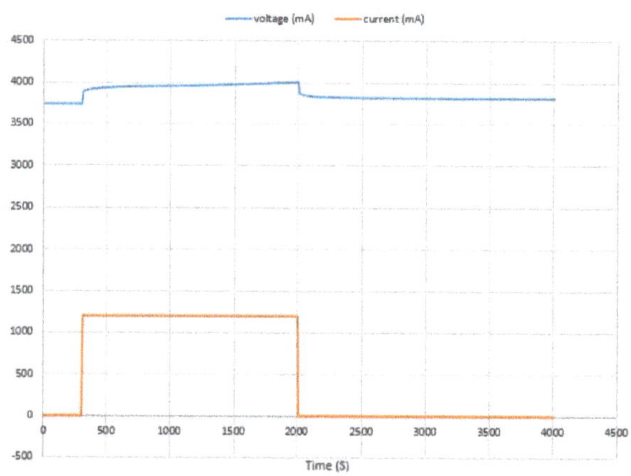

Figure 10. Voltage response of the battery to charging current pulse.

6. Conclusions

This work reported the progress of designing and implementing an automated battery test system. The tests performed generated acceptable results compared to results reported in the literature. The reading voltage and current error were around 0.5 m, which validates the work and makes the obtained results encouraging to carry on the implementation of a customized battery characterization system. However, work is still needed to complete the temperature-controlled chamber and the software running the overall system. In addition, the discharging and charging circuit need to be improved for more stability, better performance, and accuracy in general.

Author Contributions: Conceptualization, A.Z., H.B. and A.M.; methodology, A.Z. and H.B.; software, A.Z.; validation, H.B., A.Z. and A.M.; formal analysis, A.Z.; investigation, A.Z. and H.B.; resources, A.Z. and H.B.; data curation, A.Z.; writing—original draft preparation, A.Z. and H.B.; writing—review and editing, A.Z., H.B. and A.M.; visualization, H.B. and A.M.; supervision, H.B. and A.M.; project administration, H.B.; funding acquisition, A.Z. All authors have read and agreed to the published version of the manuscript.

Funding: This research received no external funding.

Institutional Review Board Statement: Not applicable.

Informed Consent Statement: Not applicable.

Data Availability Statement: Data is unavailable.

Conflicts of Interest: The authors declare no conflict of interest.

References

1. Zhou, W.; Zheng, Y.; Pan, Z.; Lu, Q. Review on the battery model and SOC estimation method. *Processes* **2021**, *9*, 1685. [CrossRef]
2. Murnane, M.; Ghazel, A. A closer look at state of charge (SOC) and state of health (SOH) estimation techniques for batteries. *Analog. Devices* **2017**, *2*, 426–436.
3. Chiasson, J.; Vairamohan, B. Estimating the state of charge of a battery. In Proceedings of the 2003 American Control Conference, Denver, CO, USA, 4–6 June 2003; Volume 4, pp. 2863–2868.
4. Vidal, C.; Malysz, P.; Kollmeyer, P.; Emadi, A. Machine learning applied to electrified vehicle battery state of charge and state of health estimation: State-of-the-art. *IEEE Access* **2020**, *8*, 52796–52814. [CrossRef]
5. Mussi, M.; Pellegrino, L.; Restelli, M.; Trovò, F. An online state of health estimation method for lithium-ion batteries based on time partitioning and data-driven model identification. *J. Energy Storage* **2022**, *55*, 105467. [CrossRef]
6. Liu, K.; Li, K.; Peng, Q.; Zhang, C. A brief review on key technologies in the battery management system of electric vehicles. *Front. Mech. Eng.* **2019**, *14*, 47–64. [CrossRef]
7. Lee, S.; Kim, J.; Lee, J.; Cho, B.H. State-of-charge and capacity estimation of lithium-ion battery using a new open-circuit voltage versus state-of-charge. *J. Power Sources* **2008**, *185*, 1367–1373. [CrossRef]
8. Zheng, F.; Xing, Y.; Jiang, J.; Sun, B.; Kim, J.; Pecht, M. Influence of different open circuit voltage tests on state of charge online estimation for lithium-ion batteries. *Appl. Energy* **2016**, *183*, 513–525. [CrossRef]
9. Ng, K.S.; Moo, C.S.; Chen, Y.P.; Hsieh, Y.C. Enhanced coulomb counting method for estimating state-of-charge and state-of-health of lithium-ion batteries. *Appl. Energy* **2009**, *86*, 1506–1511. [CrossRef]
10. Rzepka, B.; Bischof, S.; Blank, T. Implementing an Extended Kalman Filter for SoC Estimation of a Li-Ion Battery with Hysteresis: A Step-by-Step Guide. *Energies* **2021**, *14*, 3733. [CrossRef]
11. He, H.; Xiong, R.; Fan, J. Evaluation of lithium-ion battery equivalent circuit models for state of charge estimation by an experimental approach. *Energies* **2011**, *4*, 582–598. [CrossRef]
12. Cuadras, A.; Kanoun, O. SoC Li-ion battery monitoring with impedance spectroscopy. In Proceedings of the 2009 6th International Multi-Conference on Systems, Signals and Devices, Djerba, Tunisia, 23–26 March 2009; pp. 1–5.
13. How, D.N.; Hannan, M.A.; Lipu, M.H.; Ker, P.J. State of charge estimation for lithium-ion batteries using model-based and data-driven methods: A review. *IEEE Access* **2019**, *7*, 136116–136136. [CrossRef]
14. Pascoe, P.E.; Anbuky, A.H. Automated battery test system. *Measurement* **2003**, *34*, 325–345. [CrossRef]
15. Alessandrini, S.; Rizzuto, E.; Del Prete, Z. Characterizing different types of lithium-ion cells with an automated measurement system. *J. Energy Storage* **2016**, *7*, 244–251. [CrossRef]
16. Imran, R.M.; Li, Q.; Flaih, F.M.F. An Enhanced Lithium-Ion Battery Model for Estimating the State of Charge and Degraded Capacity Using an Optimized Extended Kalman Filter. *IEEE Access* **2020**, *8*, 208322–208336. [CrossRef]

engineering proceedings

Proceeding Paper

Optimization of a Speed Controller of a DFIM with Metaheuristic Algorithms †

Chetioui Lotfi [1,*], Zennir Youcef [2], Arabi Marwa [2], Horst Schulte [3], Bendib Riad [2] and Mechhoud El-Arkam [2]

1 LRPCSI Laboratory, Université 20 Aout 1955, Skikda 21000, Algeria
2 Automatic Laboratory of Skikda (LAS), Université 20 Aout 1955, Skikda 21000, Algeria
3 Control Engineering Group, Department of Engineering, University of Applied Sciences, 10115 Berlin, Germany
* Correspondence: l.chetioui@univ-skikda.dz
† Presented at the 2nd International Conference on Computational Engineering and Intelligent Systems, Online, 18–20 November 2022.

Abstract: This study proposes a control architecture based on iterative stochastic algorithms to control the doubly-fed induction machine (DFIM), often used in wind systems. We briefly present the modeling of this machine. We describe in more detail the control architecture based on indirect vector control by flux orientation (IFOC). This control architecture is optimized with the application of iterative stochastic algorithms applied to the speed controller. The optimization algorithms used were Particle Swarm Optimization (PSO), Harris Hawk Optimization (HHO) and Salp Swarm Algorithm (SSA). A comparative study under Matlab/Simulink between the different results obtained clearly shows the effectiveness of our approach.

Keywords: doubly-fed induction machine; IFOC-PI controller; meta-heuristic algorithms

Citation: Lotfi, C.; Youcef, Z.; Marwa, A.; Schulte, H.; Riad, B.; El-Arkam, M. Optimization of a Speed Controller of a DFIM with Metaheuristic Algorithms. *Eng. Proc.* **2023**, *29*, 13. https://doi.org/10.3390/engproc2023029013

Academic Editors: Abdelmadjid Recioui, Hamid Bentarzi and Fatma Zohra Dekhandji

Published: 16 January 2023

1. Introduction

In the field of electrical energy, the world is trying to reduce its dependence on fossil fuels, which are characterized by both high greenhouse gas emissions and volatile prices. The electricity industry strives to diversify its energy mix, especially with clean and renewable energy from geothermal, biomass, solar and wind [1]. One of the current research priorities is the generation of electrical energy with the help of doubly-fed asynchronous machines using wind energy as the driving force. Integrated into a wind power system, the DFIM enables operation over a wide range of wind speeds and obtaining the maximum possible power for each wind speed. Its stator circuit is directly connected to the power grid. A second circuit on the rotor is also connected to the grid, but via power converters [2]. Because less rotor power is transferred, the cost of the converters is reduced compared to a variable speed wind turbine powered by converters on the stator. This is the main reason why this machine is used for high power generation. A second reason is the possibility of adjusting the voltage at the connection point of this machine.

The aim of this work is to improve the rotor speed control of the DFIM machine by using different meta-heuristic optimization algorithms. To achieve this goal, the work is organized as follows: an introduction and then a mathematical model of the whole, then the vector control, the optimization of the parameters of the machine's speed controller with the meta-heuristic algorithms (PSO, HHO, SSA) of the simulation results in the Matlab/Simulink environment and finally a conclusion.

2. Modeling Machine and Control Design

The model of our system (DFIM) in the *d*-*q* reference [3] is as follows:

$$\mathrm{V_{sd}} = \mathrm{R_s i_{sd}} + \frac{\mathrm{d}}{\mathrm{dt}}\varphi_{\mathrm{sd}} - \mathrm{w_s}\varphi_{\mathrm{sq}} \tag{1}$$

$$V_{sq} = R_s i_{sq} + \frac{d}{dt}\varphi_{sq} - w_s\varphi_{sd} \tag{2}$$

$$V_{rd} = R_r i_{rd} + \frac{d}{dt}\varphi_{rd} - (w_s - w_r)\varphi_{rq} \tag{3}$$

$$V_{rq} = R_r i_{rq} + \frac{d}{dt}\varphi_{rq} - (w_s - w_r)\varphi_{rd} \tag{4}$$

$$\varphi_{sd} = L_s i_{sd} + L_m I_{rd} \tag{5}$$

$$\varphi_{sq} = L_s i_{sq} + L_m I_{rq} \tag{6}$$

$$\varphi_{rd} = L_r i_{rd} + L_m I_{sq} \tag{7}$$

$$\varphi_{rq} = L_r i_{rd} + L_m I_{sq} \tag{8}$$

$$P_s = \frac{3}{2}(V_{sd} i_{sd} + V_{sq} i_{sq}) \tag{9}$$

$$Q_s = \frac{3}{2}(V_{sq} i_{sd} - V_{sd} i_{sq}) \tag{10}$$

$$P_r = \frac{3}{2}(V_{rd} i_{rd} + V_{rq} i_{rq}) \tag{11}$$

$$Q_r = \frac{3}{2}(V_{rq} i_{rd} - V_{rd}) \tag{12}$$

$$C_{em} = \frac{3}{2} p \frac{L_m}{L_s}(\varphi_{qs} i_{dr} - \varphi_{ds} i_{qr}) \tag{13}$$

We applied an IFOC control method [4] to the DFIM and the Figure 1 illustrates the block diagram of control and in Figures 2 and 3, we illustrated the control diagram of current I_s and speed Ω respectively.

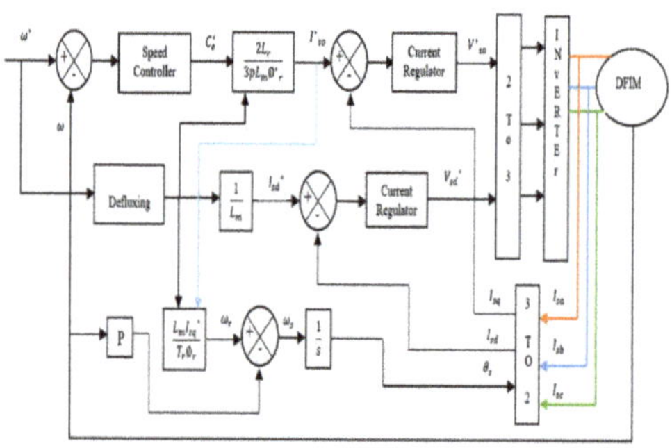

Figure 1. Block diagram of control.

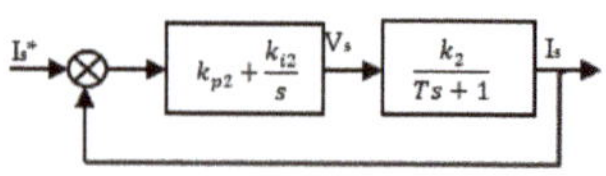

Figure 2. Control diagram of Is.

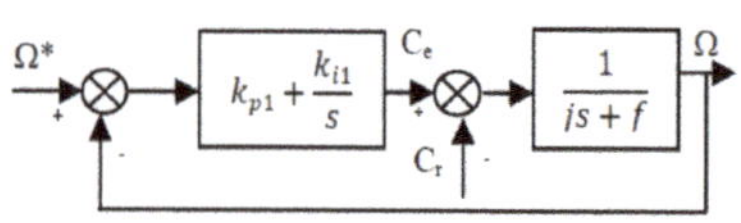

Figure 3. Control diagram of speed.

With I_s is I_{sd} or I_{sq} and V_s is V_{sd} or V_{sq}.

With:

$$k_2 = \frac{1}{R_s + \left(\frac{L_m^2}{L_r}\right)R_r}, = \frac{\sigma L_s}{R_s + \left(\frac{L_m^2}{L_r}\right)R_r}, k_{p2} = \frac{2\sigma\omega_0 T - 1}{K_2}, k_{i2} = \frac{\omega_0{}^2 T}{K_2}, k_{p1} = \frac{4j}{Ts}, k_{i1} = \frac{4j}{Ts^2}, \Omega: \text{ speed.}$$

3. Proposed Optimization Algorithm

The optimization with iterative stochastic algorithms is an approach to finding the best solution to complex problems by minimizing one or more fitness functions [5]. These techniques are applied in various engineering fields. In the literature, we have different methods of optimization such as combinatorial optimization, nonlinear programming and linear programming. These methods are designed to solve a wide range of optimization engineering problems [6]. The heuristics and meta-heuristics optimization techniques are classified as exact methods and approximation or stochastic methods. Exact methods are used for combinatorial optimization problems [7]. They are also classified according to the formulation of the graphs, i.e., the arc-node formulation, the path formulation, the arc-based formulation and the spanning tree formulation [8]. The other optimization methods based on approximate methods can be divided into two types: heuristic algorithms and meta-heuristic algorithms. Heuristic algorithms obtain their approximate solutions with a certain acceptable temporal and spatial complexity. In the literature [9], there are different heuristic strategies successfully applied to many problems, such as Genetic Algorithm (GA), Ant Colony Optimization (ACO), Simulated Annealing (SA), Particle Swarm Optimization (PSO), Gray Wolf Optimization (GWO), etc. [9].

3.1. Particle Swarm Optimization (PSO)

This optimization algorithm is based on a stochastic evolutionary calculation technique. The basic version of this algorithm is inspired by phenomenal biology from research on swarms such as schools of fish and flocks of birds [10–16]. Another modified version of this algorithm was introduced in 1998 to improve the performance of the basic version. In the principle of function of the PSO algorithm, the social behavior is modeled by a mathematical equation making it possible to guide the particles during their process of displacement [11–16]. The movement of a particle is influenced by three parameters: the inertia component, the cognitive component and the social component. Each parameter reflects a part of the equation, Figure 4 represents the flowchart of the algorithm.

3.2. Harris Hawks Optimization (HHO)

The principle of The Harris Hawk Optimization algorithm (HHO) is inspired by cooperative hunting behavior. This principle, which is based on the population and which is inspired by nature, has been modeled by a mathematical model [12–17]. Its principle goes through three phases: the exploration phase, the phase of the transition from exploration to exploitation and the last phase is the actual exploitation. A flowchart representing the principle of this algorithm is illustrated in the following figure (Figure 5). The first phase of the algorithm is the exploration phase. This phase of the algorithm mimics the search behavior of hawks for prey. The mathematical model of this step is illustrated by the following equations [13–19]:

$$x(\tau+1) = \begin{cases} x_r(\tau) - r_1|x_r(\tau) - 2r_2x(\tau)| fork \geq 0.5 \\ (x_p(\tau) - x_m(\tau)) - r_3(L_B + r_4(u_B - L_B)) fork < 0.5 \end{cases} \quad (14)$$

where K, r_1, r_2, r_3 and r_4 are random parameters. $x_p(\tau)$ stands for the prey's position whereas the current and the next positions of the hawk are shown by $x(\tau)$ and $x(\tau+1)$, respectively. The random parameters are updated in each iteration. U_B and L_B are the search space bounds. $x_r(\tau)$ refers to a randomly selected hawk and x_m refers to the average position of the population of hawks.

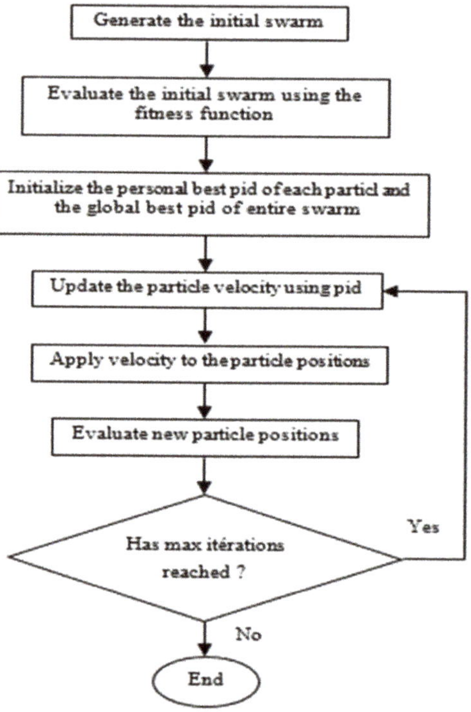

Figure 4. Flowchart of PSO.

3.3. Salp Swarm Algorithm (SSA)

The SSA algorithm simulates the swarming principle of salps when they are feeding in the oceans [14]. Salps usually form a swarm called a salp chain, the leader of the group is the salp at the top of the chain and any remaining salps are called followers. This algorithm follows the same principle of the other algorithms, i.e., the position of the salps is defined in a search space of s dimensions, where s is the number of variables in a given problem. Hence, the position of all salps is stored in a two-dimensional matrix called z. We also assumed that there was a food source called P in the search space and that this was the swarm's target. The mathematical model that models the principle of this algorithm is given as follows: The head salp can change position using the following equation [11–14]:

$$z_n^1 = \begin{cases} P_n + r_1((U_n - l_n)r_2 + l_n) & r_3 \geq 0 \\ P_n - r_1((U_n - l_n)r_2 + l_n) & r_3 < 0 \end{cases} \tag{15}$$

where the meanings of all symbols are shown in Table 1.

$$r_1 = 2e^{-}\left(-\frac{4a}{A}\right)^2 \tag{16}$$

The coefficient r_1 provides a balance between exploration and exploitation capabilities. The variation of the positions of the followers is modeled by the following equation [18,19]:

$$z_n^m = \frac{1}{2}ce^2 + v_0 e \tag{17}$$

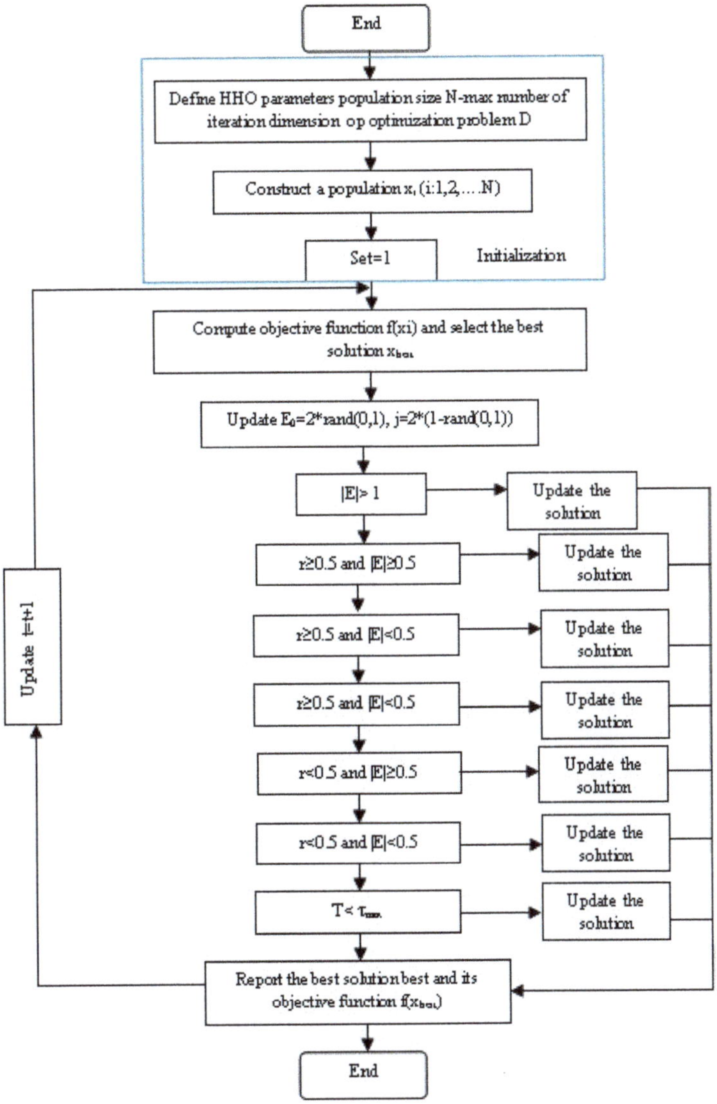

Figure 5. Flowchart of HHO.

Table 1. The meanings of all symbols.

Symbols	Meaning
z_n^1	leader position
P_n	food source position
U_n	upper bound of
l_n	lower bound of
r_1, r_2, and r_3	Random variables in [0, 1].

Table 1. *Cont.*

Symbols	Meaning
a	current iteration
A	maximum number of iterations
z_n^m	position of mith follower salp
e	Time
v_0	Initial speed

The flowchart of this algorithm is given as follows figure (Figure 6):

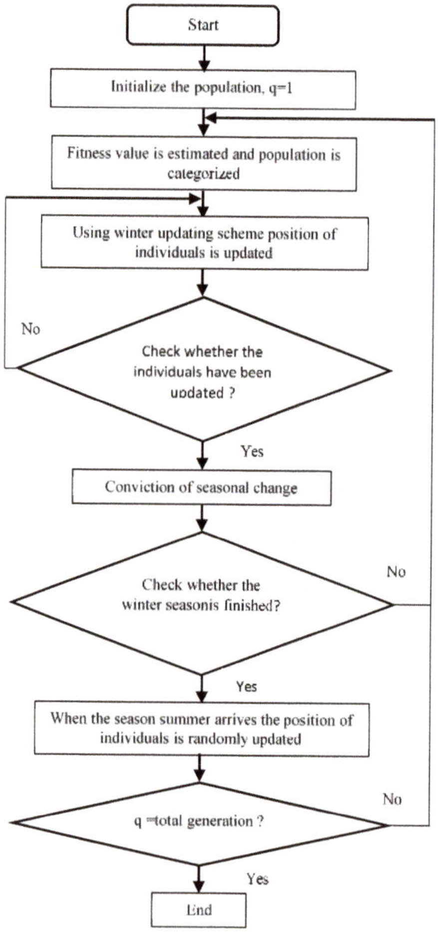

Figure 6. Flowchart of SSA algorithm.

4. Results and Discussions

The simulation was carried out under MATLAB/SIMULINK. We present the results obtained with the control architecture chosen to control our system (DFIM). We used the different algorithms (PSO, SSA and HHO) to optimize the speed controller parameters [13–15] of the machine shown in Figure 7. In order to obtain the optimal solution, there are several formulas of the object function. Among the formulas (performance indices), we used the following four formulas: square integral error (ISE), absolute integral error (IAE), Absolute Temporal Integral Error (ITAE) and Square Temporal Integral Error (ITSE). They have defined themselves as follows:

$$\mathrm{IAE} = \int |e(t)|dt \quad (18)$$

$$\mathrm{ITAE} = \int t.|e(t)|dt \quad (19)$$

$$\mathrm{ISE} = \int e(t)^2 dt \quad (20)$$

$$\mathrm{ITSE} = \int t.\left|e(t)^2\right|dt \quad (21)$$

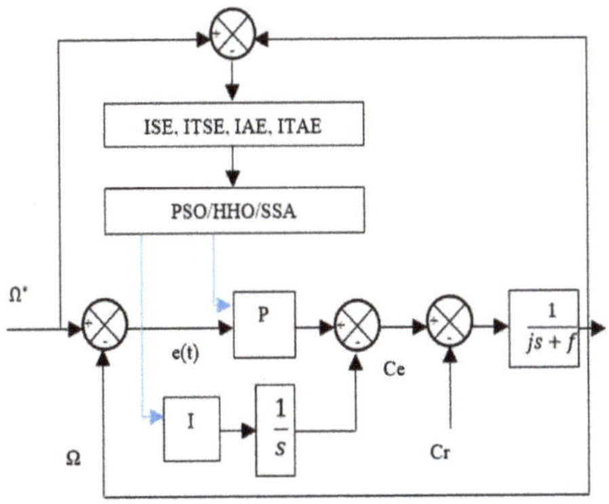

Figure 7. The control block diagram of speed.

The different simulation results obtained with this algorithm are illustrated in the following table (Table 2):

Table 2. The different results obtained by optimization.

Algorithms	ISE	ITSE	IAE	ITAE	Error
PSO-Kp	100	0.0001	1.4932	23.2891	0.00004
PSO-Ki	100	100	81.8677	30.2755	
HHO-Kp	100	100	100	100	−0.0002
HHO-Ki	98.1475	0.1499	36.6452	0.008	
SSA-Kp	10	9.4731	9.8906	8.3202	−0.04091
SSA-Ki	0.4647	0.3370	0.3370	0.2921	

The values colored in yellow in each function are the best for a minimum error.

The curves below represent the characteristics of the machine, such as torque, current and speed, with the different algorithms applied.

The curves presented represent the results obtained by the various algorithms applied with a resisting torque (Cr = 0 (Nm)). In Figure 8, we note that the torque is stabilized at a value equal to 0.1 (s) and for the currents shown in Figures 9 and 10, we note that the currents are stable from 0.08 s. According to the speed curve obtained by the different algorithms applied (Figure 11), we can observe that the speed curve with the PSO algorithm (Figure 12) is the most efficient for our system, with a minimum error between the machine speed and the reference speed of 4×10^{-4}. We can also note a very good rise time of 1.4358 s. for Figure 13. We plotted the speed curve with a resistance torque of 100 (Nm) against 1000 (s) to see the impact of a disturbance on our regulation, noting that the speed curve with the parameters optimized by the algorithm (PSO) quickly returned to its input speed and was less disturbed than the output speed optimized with (HHO) and (SSA).

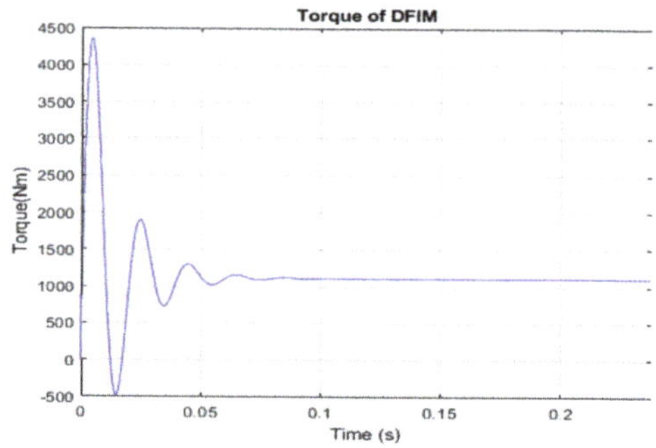

Figure 8. The curve of electromagnetic.

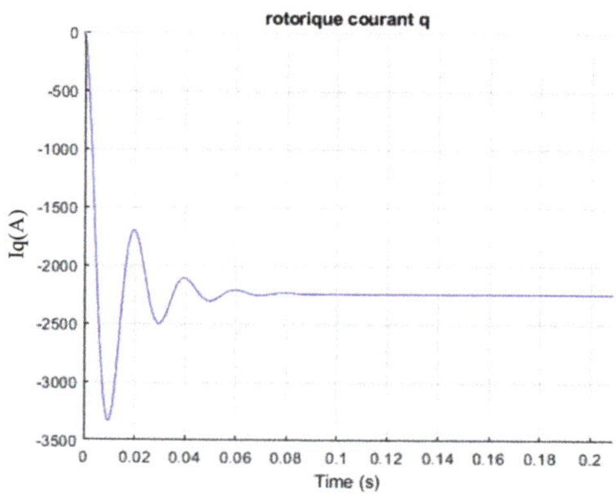

Figure 9. The rotor current on the q axis.

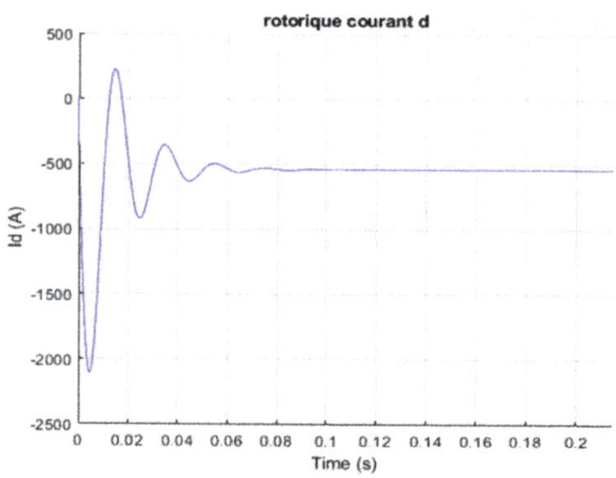

Figure 10. The rotor current on the d axis.

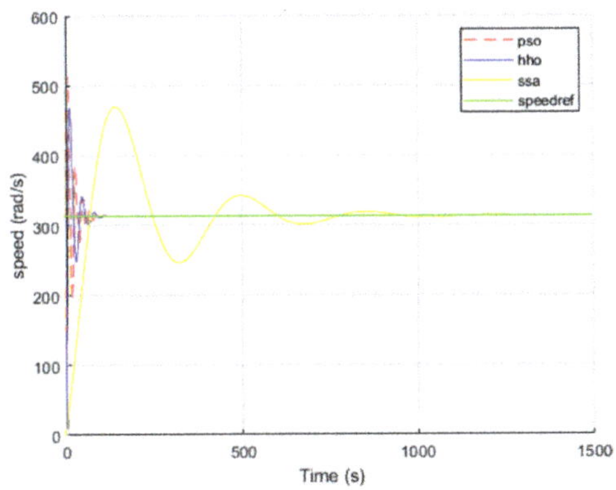

Figure 11. The mechanical speed.

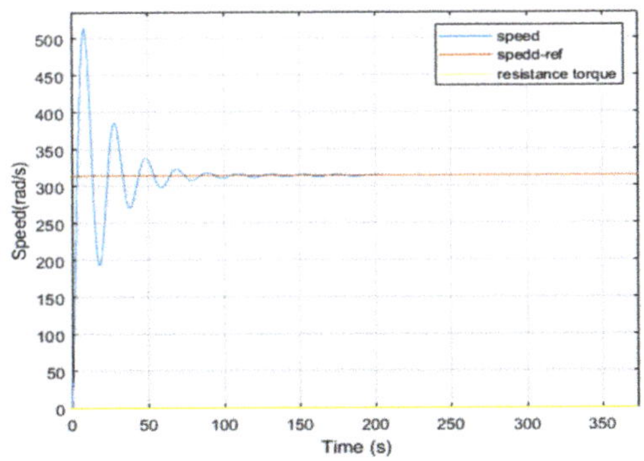

Figure 12. The mechanical speed with PSO.

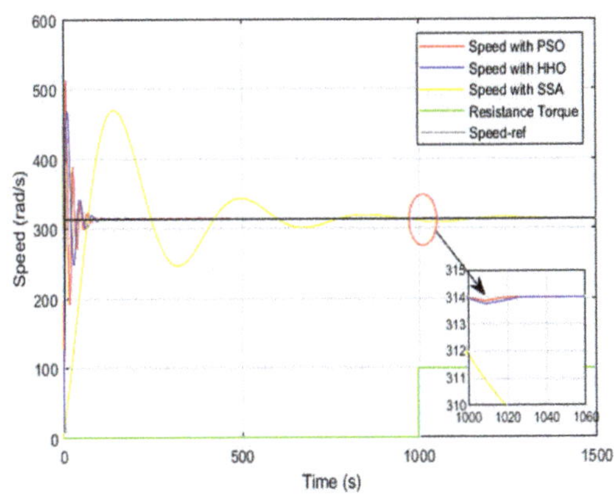

Figure 13. The mechanical speed with resistance torque.

5. Conclusions

In this study, we proposed a control architecture based on meta-heuristic algorithms to control the speed of our system (DFIM), since it is an important parameter to maintain a good energy rounding used in the renewable energies. To obtain optimal PID controller parameters, the meta-heuristic algorithms known as Harris Hawks Optimization (HHO),

Particle Swarm Optimization (PSO) and Slap Swarm (SSA) were used. The comparison between the three algorithms used with the analysis of the fitness function was the speed error in our study. The results show that the proposed PSO-based approach can improve the stability performance and minimize the speed control error. In future work, we will apply our control design for the entire wind turbine conversion.

Author Contributions: Conceptualization, C.L. and Z.Y.; methodology, C.L.; software, A.M.; validation, Z.Y. and H.S.; formal analysis, C.L., A.M. and Z.Y.; writing—original draft preparation, C.L., A.M., Z.Y., M.E.-A., B.R. and H.S.; writing—review, C.L., Z.Y. All authors have read and agreed to the published version of the manuscript.

Funding: This research received no external funding.

Data Availability Statement: Not applicable.

Conflicts of Interest: The authors declare no conflict of interest.

References

1. Fatima, M.; Assia, H.; Habib, H. Commande d'un système éolien basé sur une MADA. *Intent. J. Sci. Res. Eng. Technol. (IJSET)* **2019**, *11*, 9–14.
2. Poitiers, F. Etude et Commande de Génératrices Asynchrones pour l'Utilisation de l'Energie Eolienne. Thèse de Doctorat, Ecole Polytechnique de l'Université de Nantes, Nantes, France, 2003; p. 167.
3. Lesenne, J.; Notelet, F.; Seguier, G. *Introduction à l'éctrotechnique approfondie*; Tech-nique et Documentation: Paris, France, 1981; p. 247.
4. Ozkan, A.K.I.N.; Irfan, A.L.A.N. The use of FPGA in field-oriented control of an induction machine. *Turk. J. Electr. Eng. Comput.* **2010**, *18*, 943–962.
5. Gavrilas, M. Heuristic and metaheuristic optimization techniques with application to power systems. In Proceedings of the 12th WSEAS international Conference on Mathematical Methods and Computational Techniques in Electrical Engineering, Timisoara, Romania, 21–23 October 2010; pp. 95–103.
6. Franz, R. *Optimization Problems*; Chapter 1; Springer: Berlin/Heidelberg, Germany, 2011; pp. 7–44.
7. Martí, R.; Reinelt, G. *The Linear Ordering Problem: Exact and Heuristic Methods in Combinatorial Optimization*; Springer Science & Business Media: New York, NY, USA, 2011; Volume 175, pp. 172–184.
8. Dixit, A.; Mishra, A.; Shukla, A. Vehicle Routing Problem with Time Windows Using Meta-Heuristic Algorithms: A Survey. In *Harmony Search and Nature Inspired Optimization Algorithms*; Springer: Singapore, 2019; pp. 539–546.
9. Yu, X.; Gen, M. *Introduction to Evolutionary Algorithms*; Springer Science & Business Media: New York, NY, USA, 2010; p. 433.
10. Ou, C.; Lin, W. Comparison between PSO and GA for parameters optimization of PID controller. In Proceedings of the IEEE International Conference on Mechatronics and Automation, Luoyang, China, 25–28 June 2006; pp. 2471–2475.
11. Kennedy, J.; Eberhart, R.C. Particle swarm optimization. In Proceedings of the IEEE International Conference on Neural Networks, Perth, WA, Australia, 27 November–1 December 1995; IEEE Service Center: Piscataway, NJ, USA, 1995; pp. 1942–1948.
12. Heidari, A.; Mirjalili, S.; Faris, H.; Aljarah, I.; Mafarja, M.; Chen, H. Harris hawks optimization: Algorithm and applications. *Futur.Gener. Comput. Syst.* **2019**, *97*, 849–872. [CrossRef]
13. Gaing, Z.-L. A particle swarm optimization approach for optimum design of PID controller in AVR system. *IEEE Trans. Energy Convers.* **2004**, *19*, 384–391. [CrossRef]
14. Hegazy, A.E.; Makhlouf, M.; Tawel, G. Improved Salp Swarm Algorithm for Feature Selection. *J. King Saud Univ. Comput. Inf. Sci.* **2020**, *32*, 335–344. [CrossRef]
15. Ahmadi, S.-A. Human behavior-based optimization: A novel metaheuristic approach to solve complex optimization problems. *Neural Comput. Appl.* **2017**, *28*, 233–244. [CrossRef]
16. Ekinci, S.; Hekimoglu, B. Improved kidney-inspired algorithm approach for tuning of PID controller in AVR system. *IEEE Access* **2019**, *7*, 39935–39947. [CrossRef]
17. Sulaiman, D.R. Multi-objective Pareto front and particle swarm optimization algorithms for power dissipation reduction in microprocessors. *Int. J. Electr. Comput. Eng. (IJECE)* **2020**, *10*, 6549–6557. [CrossRef]
18. Izci, D.; Ekinci, S.; Demiroren, A.; Hedley, J. HHO Algorithm based PID Controller Design for Aircraft Pitch Angle Control System. In Proceedings of the International Congress on Human-Computer Interaction, Optimization and Robotic Applications (HORA), Ankara, Turkey, 26–28 June 2020; p. 6.
19. Kose, E. Optimal Control of AVR System with Tree Seed Algorithm-Based PID Controller. *IEEE Access* **2020**, *8*, 89457–89467. [CrossRef]

engineering proceedings

Proceeding Paper

High Gain Observer Based Active Disturbance Estimation ADE for Second Order Nonlinear Uncertain Systems (ex:Induction Motor) †

Saad Kelam [1,*], Mohamed Chennafa [2], Mohamed Belkheiri [3] and Mohamed Amine Zaafrane [4]

1 Automatic Departement, University of Sciences and Technology of Oran—Mohamed-Boudiaf (USTO-MB), El Mnaouar, BP 1505, Bir El Djir 31000, Algeria
2 National Polytechnic School of Oran-Maurice Audin, Route d'Es-Sénia, B.P. 1523, El M'Naouer, Oran 31000, Algeria
3 The Telecommunications, Signals and Systems Laboratory, University Amar Telidji of Laghouat, B.P. 37G, Guardaia Road, Laghouat 03000, Algeria
4 Electrical Engineering Faculty, University of Sciences and Technology of Oran—Mohamed-Boudiaf (USTO-MB), El Mnaouar, BP 1505, Bir El Djir 31000, Algeria
* Correspondence: saad.kelam@univ-usto.dz
† Presented at the 2nd International Conference on Computational Engineering and Intelligent Systems, Online, 18–20 November 2022.

Abstract: The main objective of this paper is to solve state observation and external disturbances estimation for a class of second-order nonlinear systems. The proposed method relies mainly on the high gain observer as an estimator that tries to estimate the state vector and at the same time identifies the system's unknown combined structured and unstructured uncertainties. The efficiency of the proposed method is demonstrated by estimating the flux and the speed of the induction motor by simulation.

Keywords: high-gain observers; estimation; disturbances; second order systems; uncertain systems; active disturbance estimation; flux; speed; induction motor

Citation: Kelam, S.; Chennafa, M.; Belkheiri, M.; Zaafrane, M.A. High Gain Observer Based Active Disturbance Estimation ADE for Second Order Nonlinear Uncertain Systems (ex:Induction Motor). *Eng. Proc.* **2023**, *29*, 14. https://doi.org/10.3390/engproc2023029014

Academic Editors: Abdelmadjid Recioui, Hamid Bentarzi and Fatma Zohra Dekhandji

Published: 17 January 2023

1. Introduction

Dynamic systems are usually exposed to control signals that can be used to drive them to achieve a given objective and solve regulation and tracking control problems based on reliable models and in the presence of external perturbations and disturbances. These disturbances are, generally speaking, unstructured, i.e., they may depend on internal variables of the system in an either unknown or an ignored manner or else, they may also depend on external phenomena affecting the system [1]. They are unknown, unpredictable, or unmodeled. They can affect the controlled system response, its stability and closed-loop performance if they are ignored in the control synthesis phase. Solving this problem is still an open issue for control systems researchers. In the literature, many control methods that rely on a reliable model that approximates the system to be controlled are proposed feedback linearization and backstepping. In this case, some analytic robust terms are included in the derivation of the controller to guarantee stability and performance in the presence of uncertainties [2,3]. Sliding mode control techniques are among the robustifying methods that derive the control law bead on two combined terms, the equivalent control term based on the system model and a switching robustifying term [4]. High-gain observers are first used in the context of linear feedback as a tool for robust observer design for loop transfer recovery achieved by state feedback and in robust H∞ control in [5]. High-gain observers integration in nonlinear feedback control started in the late 1980 [6–9], where researchers in these papers have studied a wide range of nonlinear control issues, including stabilization, regulation, tracking, and adaptive control. They also studied time-varying high-gain observers and related problems [5]. In any case, disturbances are classed into two main categories and referred to as internal uncertainties and disturbances [10–14]. The main idea of this short note is to simplify the prediction of disturbances acting on the plant by grouping all external and internal dynamic disturbances into a single disturbance term and proceed to estimate the effects of this disturbance using High-Gain observers. The rest of the paper is organized as follows: in Section 2, we present the problem statement and state observation scheme with active disturbance estimation (ADE) based on the high-gain observer. The resulting ADE observer is applied to the induction motor system in Section 3 followed by the concluding remarks in Section 4.

2. State Observation Scheme with Active Disturbance Estimation (ADE)

2.1. Problem Statement

Active Disturbance Estimation ADE strategy is based on the possibility of online estimating of unknown disturbance inputs affecting the plant. The feedback control law usually requires full knowledge of the state which can be at the same time reconstructed by the same designed observer. In this section, we consider the following endogenously perturbed second-order system:

$$\begin{cases} \dot{x}_1(t) = x_2(t) \\ \dot{x}_2(t) = u(t) + \theta^T \phi(x(t)) + \varepsilon(t) \end{cases} \quad (1)$$

where $x(t) = [(x_1(t)\&x_2(t))]^T$ is the state vector and the first element $x_1(t)$ of $x(t)$ is only the measured state or output $y(t) = x_1(t)$ and $\theta^T \phi(x(t))$ known linearly parameterized terms and $\varepsilon(t)$ represents the unmodelled unknown uncertain terms. The system defined in (1) can be used to derive an adaptive high-gain observer that reconstructs the state vector with an adaptive law that estimates the unknown parameter vector θ the observer error is bounded and is ensured to converge given some robustness conditions on the term $\varepsilon(t)$ to be verified. For more information on designing high-gain observers, the readers can refer to reference [5]. In the context of Active Disturbance Estimation ADE disturbances and nonlinearities may be combined in a single term that can be approximated online. This suggests rewriting the second-order uncertain system in the following form:

This is an example of the equation:

$$\begin{cases} \dot{x}_1(t) = x_2(t) \\ \dot{x}_2(t) = u(t) + \xi(t) \end{cases} \quad (2)$$

where $\xi(t) = \theta^T \phi(x(t)) + \varepsilon(t)$.

The term $\xi(t)$ combines all the unknown exogenous terms. It is assumed that $\xi(t)$ and its first-order time derivative are uniformly absolutely bounded. Note that many mechanical electrically driven systems in robotics and mechatronics can be represented by the system in Equation (1). The objective can then be stated as follows: to design a state observer to estimate the state vector and actively estimates the unknown exogenous terms with converging observation errors.

2.2. High-Gain Observer with Active Disturbance Estimator

Consider the following Extended State Observer (ADE):

$$\begin{cases} \dot{\hat{x}}_1(t) = \hat{x}_2 + \lambda_1(x_1(t) - \hat{x}_1(t)) \\ \dot{\hat{x}}_2(t) = u(t) + \hat{\xi}(t) + \lambda_2(x_1(t) - \hat{x}_1(t)) \\ \dot{\hat{x}}_3(t) = \lambda_3(x_1(t) - \hat{x}_1(t)) \end{cases} \quad (3)$$

where λ_1,λ_2 and $\lambda 3$ are (ADE) observer gains and they are chosen to achieve convergence.

$$\lambda_1 = \frac{1}{\varepsilon} \times (p_0 + 2\xi_0 \omega_0)$$

$$\lambda_2 = \frac{1}{\varepsilon^2} \times (2p_0 \xi_0 \omega_0 + \omega_0^2)$$

$$\lambda_3 = \frac{1}{\varepsilon^2} \times (p_0\omega_0^2)$$

with p_0, ξ_0,ω_0 are positive constants and ε being a small parameter bestowing a high-gain character to the ESO design. Here, the variable $\hat{\xi}$ is supposed to estimate the total disturbance input to the system $\xi(t)$, composed of the two terms; the endogenous perturbation input $\theta^T\phi(x(t))$ and the exogenous perturbation input $\varepsilon(t)$.

Defining the estimation error vector as $e_1(t) = x_1(t) - \hat{x}_1(t)$,$e_2(t) = x_2(t) - \hat{x}_2(t)$ and $e_\xi(t) = \xi(t) - \hat{\xi}(t)$, its dynamics evolve according to the linear perturbed dynamics

$$\begin{cases} \dot{e}_1(t) = e_2 - \lambda_1 e_1(t) \\ \dot{e}_2(t) = e_\xi(t) - \lambda_1 e_1(t) \\ \dot{e}_3(t) = \dot{\xi} - \lambda_3(x_1(t) - \hat{x}_1(t)) \end{cases} \tag{4}$$

The derivative term $\dot{\xi}$ is given by:

$$\dot{\xi} = \theta^T \frac{d(\phi(x(t))}{dt} + \frac{d(\varepsilon(t))}{dt}$$

It is bounded given that the nonlinear terms are continuous and their derivatives are continuous as well. Another methodology based on adaptive control can be used to estimate the term $\theta^T(\phi(x(t))$ rather than $\hat{\xi}(t)$ Defining the error vector as $e^T = (e_1, e_2, e_\xi)$ Equation (4) can be written as

$$\dot{e} = \begin{pmatrix} -\lambda_1 & 1 & 0 \\ -\lambda_2 & 0 & 1 \\ -\lambda_3 & 0 & 0 \end{pmatrix} + \begin{pmatrix} 0 \\ 0 \\ 1 \end{pmatrix} \dot{\xi}(t) \tag{5}$$

The right choice of the design parameters λ_i ensure the convergence of the observer error dynamics in (4) for more details on the convergence readers may refer to the book written by Hassan Khalil on High gain observers (5).

3. Ade Observer for the Induction Motor

3.1. Induction Motor System Modeling

The induction motor is one of the most complex dynamical systems due it is difficult to control since this system is considered to be multivariate having properties such as high coupling and high non-linearity. It is generally described by a fifth-order nonlinear differential equation with two inputs. Moreover, the control and the parameters estimation of an induction motor is very complex, because it is subject to unknown disturbances and the variation of motor parameters due to heating and magnetic saturation. We used field-oriented control (FOC) to perform a change of variables to bring the equations into new coordinates that will be simple to work with, where the currents regulating the flux and the speed are decoupled [15,16]. Thus, instead of working with (ψ_{ra},ψ_{rb}), one uses the polar coordinate representation (ρ,ψ_d) [15,16]. IM is represented by the model [15–17].

$$\begin{aligned} \frac{d\omega}{dt} &= \mu\psi_d i_q - \frac{f}{J}\omega - \frac{1}{J}\tau_L \\ \frac{\psi_d}{dt} &= -\eta_{rn}\psi_d + \eta_{rn}L_m i_d \\ \frac{di_d}{dt} &= -\gamma_n i_d + \eta r_n \beta\psi_d + n_p\omega i_d + \frac{\eta_{rn}L_m i_d^2}{\psi_d} \\ &\quad + f_1 + \frac{1}{\sigma L_s}u_d + \delta_1 \\ \frac{di_q}{dt} &= -\gamma_n i_q - \beta n_p\omega\psi_d - n_p\omega i_d - \frac{\eta_{rn}L_m i_d i_q}{\psi_d} \\ &\quad + f_2 + \frac{1}{\sigma L_s}u_q + \delta_2 \\ \frac{d\rho}{dt} &= n_p\omega + \frac{\eta_{rn}L_m i_q}{\psi_d} \\ \tau_e &= J\mu\psi_d i_q \end{aligned} \tag{6}$$

In the above model, the angular speed of the rotor is denoted by ω, ψ_r is the flux in the stator reference frame, and i_S and u_S denote the stator currents and voltages, n_p is the number of pole pairs, R_s and R_r are the stator and rotor resistances, M is the mutual inductance, L_s and L_r are the stator and rotor inductances, and the two mechanical parameters: J is the inertia of the rotor and f is the load torque. The resistances R_s, R_r and the inductances L_s , L_r will be treated as uncertain parameters with R_{sn}, R_{rn} and L_{sn}, L_{rn} as their rated values, respectively. $\eta_{rn} = \frac{R_{rn}}{L_{rn}}$, $\eta_{sn} = \frac{R_{sn}}{L_{sn}}$ $\delta_1 = \delta_R r g_1 + \delta_{Sr} g_2 + \delta_{Lr} g_3 + \delta_{Ls} g_4$, $\delta 2 = \delta_R r g_5 + \delta_{Sr} g_6 + \delta_{Lr} g_7 + \delta_{Ls} g_8$. Where f_1 and f_2 are continuous functions of δ_{Rr}, δ_{Rs}, δ_{Lr}, δ_{Ls} are continuous functions of (ψ_d, i_d, i_q). The electromagnetic torque $\tau_e = J\mu\psi_d i_q$ is now just proportional to the product of two state variables ψ_d and i_q. That is the first four equations of (6) may be written as two decoupled subsystems consisting of the flux subsystem model:

$$\begin{cases} \frac{d\psi_d}{dt} = -\eta_{rn}\psi_d + \eta_{rn} M i_d \\ \frac{di_d}{dt} = f_d + u_d \end{cases} \tag{7}$$

and the speed subsystem model:

$$\begin{cases} \frac{d\omega}{dt} = \mu i_q \psi_d - \frac{f}{J}\omega - \frac{\tau_L}{J} \\ \frac{di_q}{dt} = f_q + u_q \end{cases} \tag{8}$$

where

$$f_d = -\gamma_n i_d + \eta_{rn}\beta\psi_d + n_p\omega i_q + \frac{\eta_{rn} M i_q^2}{\psi_d} + f_1 + \delta_1$$

and

$$f_q = -\gamma_n i_q - n_p\omega\beta\psi_d + n_p\omega i_q - \frac{\eta_{rn} i_d i_q}{\psi_d} + f_2 + \delta_2$$

f_d and f_q are perturbation terms. The field-oriented control consists of using u_d to force ψ_d to track the constant flux reference $\psi_{dn} = M i_{dn}$ in the flux subsystem, and the control of speed in the subsystem is done through the input u_q. Consequently, the flux dynamics are now decoupled from the speed dynamics. However, the differential equations for i_d and i_q still contain quite complicated nonlinearities in both flux (7) and speed subsystems (8). To solve the problem of unknown variations in plant parameters and structure, in this paper a robust ADE observer will be designed to eliminate the effect of unstructured uncertainties in each subsystem of the decoupled dynamics of flux and speed subsystems of the IM .

3.2. *High-Gain Observer with Active Disturbance Estimator*

Noting that in applying the feedback (6), there is some uncertainty in the knowledge of the motor parameters and the state variables. Furthermore, the motor parameters R_r and R_s can vary significantly due to Ohmic heating while L_r and L_s can also vary due to magnetic saturation [15]. For that, assuming that all neglected terms for each subsystem as an error signal Δ_i (Δ_d and Δ_q) consequently, the dynamics of the output $(y_d = x_1 = \psi_d)$ $(y_q = \xi_1 = \omega)$. Defined by (7) and (8), respectively, will be expressed as:

Flux expressed:

$$\begin{cases} \dot{x}_1(t) = x_2(t) \\ \dot{x}_2(t) = f_d(t) + u_d(t) \end{cases} \tag{9}$$

Speed expressed:

$$\begin{cases} \dot{\xi}_1(t) = \xi_2(t) \\ \dot{\xi}_2(t) = f_q(t) + u_q(t) \end{cases} \tag{10}$$

Our objective is to develop a high-gain observer with an active disturbance estimator that is capable of estimating the flux and speed of an induction motor and eliminating the uncertainties effect in the observer law in order to achieve a good estimation of the desired trajectory and to solve the problem of unknown variations in plant parameters and the load torque. In designing an observer law for speed estimation and torque load generation based on a given flux reference signal, for each subsystem, a simpler strategy is followed in this paper where two ADE observers are used to overcome the effects of nonlinear uncertainties and neglected terms in the accuracy of the estimation. The desired torque $\tau_e^* = \mu\psi_d^* i_q$ to be generated with the corresponding reference flux ψ_d^*. In order to the flux tracking, we introduce the state variables $(x_1 = \psi_d)$, $(x_2 = \dot{x}_1)$ and $(\zeta_d = f_d)$ the state space model (6) and for the tracking problem of the speed we define the new state vector $(\xi_1 = \omega)$, $(\xi_2 = \dot{\xi}_1)$ and $(\zeta_q = f_\omega)$ the speed subsystem dynamics given in (9) can be written as follows:

$$\begin{cases} \dot{x}_1(t) = x_2(t) \\ \dot{x}_2(t) = \zeta_d(t) + u(t) \\ \dot{\zeta}_d(t) = \dot{f}_d(t) \end{cases} \tag{11}$$

where

$$\zeta_d(t) = f_d(t) = \eta_{rn}^2 \psi_d - \eta_{rn}^2 + \eta_{rn} M F_d, b_d = \frac{\eta_{rn} M}{\sigma L_s}$$

and

$$u(t) = u_d(t)$$

$$\begin{cases} \dot{\xi}_1(t) = \xi_2(t) \\ \dot{\xi}_2(t) = \zeta_q(t) + u(t) \\ \dot{\zeta}_q(t) = \dot{f}_q(t) \end{cases} \tag{12}$$

where

$$\begin{aligned} f_q(t) &= \mu\psi_d\left(-\gamma_n i_q - \beta n_p \omega \varphi_d - n_p \omega i_d - \tfrac{\eta_{rn} L_m i_d i_q}{\varphi_d}\right) \\ &-\tfrac{f}{J}\dot{\omega} - \tfrac{\tau_L}{J} \end{aligned}$$

and

$$u(t) = u_q(t)$$

Note that the objective of the ADE strategy is to obtain proper estimation, minimize the total disturbance and obtain a good estimation of the flux and the speed. The ADE observer law is given by (7) and (8), where λ_i are design parameters chosen as described in Section 2. The observer uses the error between the actual measured flux ψ and the estimated flux $\hat{\psi}$ for correcting the estimated vector.

4. Results

The performance of the high-gain observer with an active disturbance estimator is tested by a differential simulation model under Matlab/Simulink environment. The simulated motor is a six-pole ($n_p = 3$), 1/12 horsepower two-phase IM and the rated parameters of the motor were taken from [15] as $R_s = 1.7\ \Omega$, $R_r = 3.9\ \Omega$, $L_s = 0.0014$ H, $L_r = 0.0014$ H, $M = 0.0117$ H, $J = 0.00011$ K·gm^2, $f = 0.00014$ N·m/rad/s.

The speed and flux estimation performance of ADE is illustrated in graphs (a) and (b) in Figure 1, respectively, which shows that the ADE observer is robust to parametric uncertainties and external disturbance variation. Although the references peed is changed from 100 rad/s to 20 rad/s and the torque load is kept constant with variations. The ADE observer has not encountered a tracking problem to follow the reference speed and flux as illustrated in Figure 1 This result shows the achieved performance of the ADE strategy for the perfect speed and flux esteeming.

For the last simulation scenario, a highly varying load torque is applied with parametric uncertainty to show the power of ADE to achieve perfect speed tracking for highly varying load illustrated in the Figure 2. Obviously, the ADE observer can estimate the speed, flux and disturbance despite a change in the torque simulation scenario as demonstrated in Figure 3.

Figure 4 illustrates the ADE observer system's U_{ABC} and d, q-axes control voltage waveform, respectively.

The external and internal disturbances are grouped in a single term and estimated by the observer it is highlighted in Figure 5. This signal can be usefull in designing feedback controllers for induction motor. The above shows the power and robustness of the observer to estimate the speed and flux in spite of the highly varying load torque and parametric uncertainty.

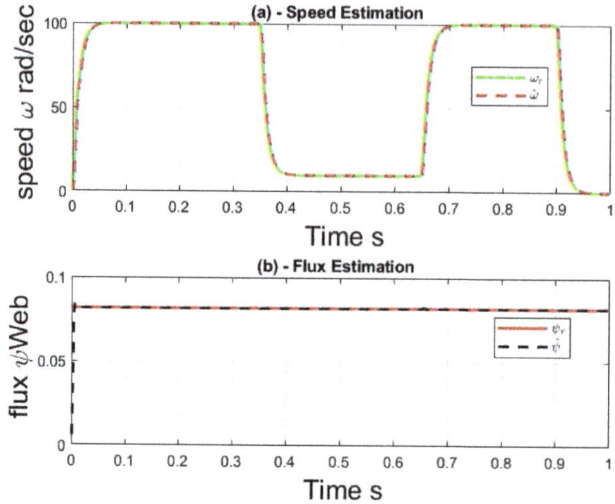

Figure 1. Simulation results with constant load and variation of parametric uncertainties with ADE Observer.

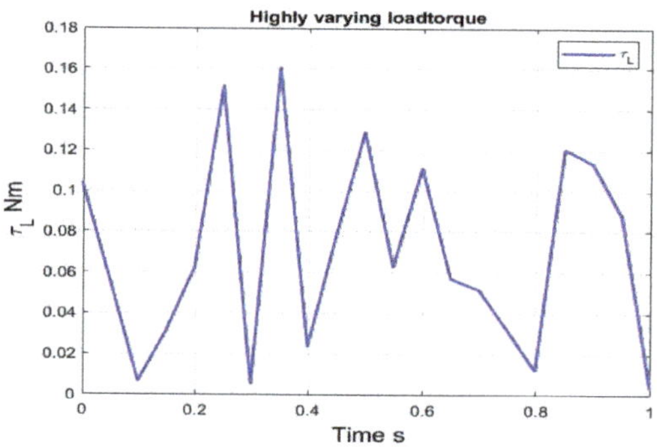

Figure 2. The applied of varying load.

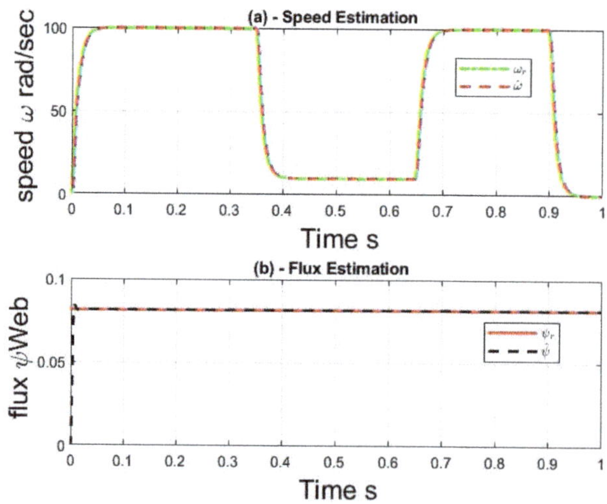

Figure 3. Simulation results with varying load and parametric uncertainties with ADE Observer.

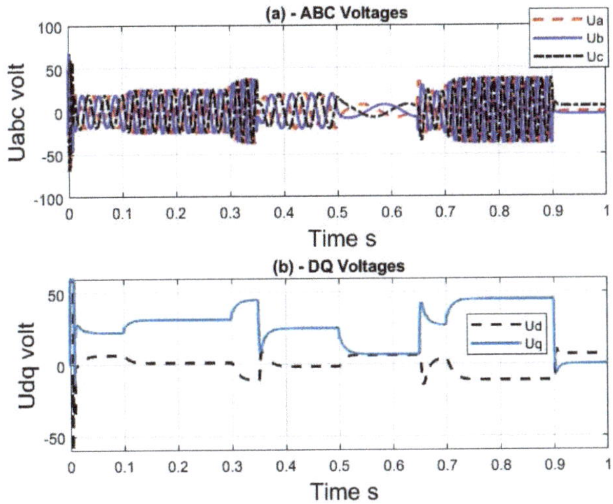

Figure 4. Simulation results with varying load and parametric uncertainties with ADE Observer.

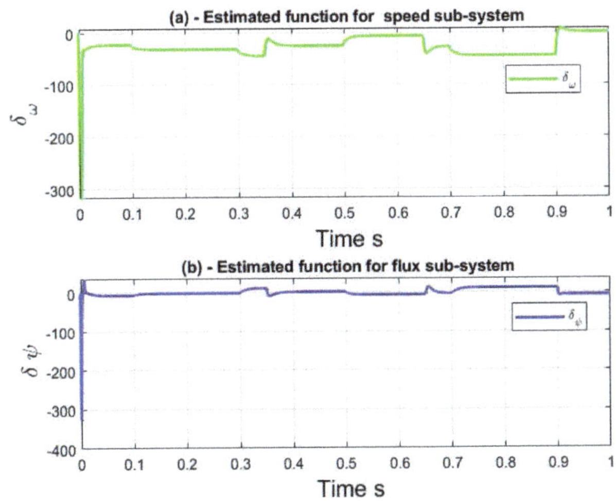

Figure 5. Simulation results with varying load and parametric uncertainties with ADE Observer.

5. Conclusions

An active disturbance estimator combined with a high gain state observer is presented in this paper to address the problem of unmodelled and external disturbances for a second-order nonlinear system. The convergence of the state estimation disturbances is guaranteed by the right choice of observer gains. We have applied this method to an induction motor to show the design choices and the achieved performance although we have assumed a partly known model of the system to be estimated. In future research, we will try to include the ADE in designing robust observers to compensate for the unknown terms.

Author Contributions: Conceptualization, S.K.; methodology, S.K.; software, S.K., M.C. and M.B.; writing—original draft preparation, S.K.; writing—review and editing, S.K., CH.M., M.B. and M.A.Z. All authors have read and agreed to the published version of the manuscript.

Funding: This work was supported by the Directorate-General for Scientific Research and Technological Development/Ministry of High Education and Scientific Research of Algeria (DGRSDT/MESRS).

Institutional Review Board Statement: Not applicable.

Informed Consent Statement: Not applicable.

Data Availability Statement: Not applicable.

Conflicts of Interest: The authors declare no conflict of interest.

Abbreviations

The following abbreviations are used in this manuscript:

FOC	Field-Oriented Control
IM	Induction motor
ADE	Active Disturbance Estimation

References

1. Sira-Ramírez, H.; Luviano-Juárez, A.; Ramírez-Neria, M.; Zurita-Bustamante, E.W. *Active Disturbance Rejection Control of Dynamic Systems: A Flatness-Based Approach*; Elsevier Ergonomics Book Series; Butterworth-Heinemann: Amsterdam, The Netherlands, 2017.
2. Ammar, A.; Kheldoun, A.; Metidji, B.; Ameid, T.; Azzoug, Y. Feedback linearization based sensorless direct torque control using stator flux MRAS-sliding mode observer for induction motor drive. *ISA Trans.* **2020**, *98*, 382–392. [CrossRef] [PubMed]
3. Regaya, C.B.; Farhani, F.; Zaafouri, A.; Chaari, A. A novel adaptive control method for induction motor based on Backstepping approach using dSpace DS 1104 control board. *Mech. Syst. Signal Process.* **2018**, *100*, 466–481. [CrossRef]
4. Regaya, C.B.; Farhani, F.; Zaafouri, A.; Chaari, A. An adaptive sliding-mode speed observer for induction motor under backstepping control. *Int. J. Innov. Comput. I* **2017**, *11*, 763–771.
5. Khalil, H.K. *High-Gain Observers in Nonlinear Feedback Control*; Society for Industrial and Applied Mathematics: Philadelphia, PA, USA, 2017.
6. Petersen, I.R.; Holot, C.V. High-gain observers applied to problems in disturbance attenuation, H-infinity optimization and the stabilization of uncertain linear systems. In Proceedings of the 1988 American Control Conference, Atlanta, GA, USA, 15–17 June 1988; pp. 2490–2496.
7. Esfandiari, F.; Khalil, H.K. Observer-based design of uncertain systems: Recovering state feedback robustness under matching conditions. In Proceedings of the 1989 American Control Conference, Pittsburgh, PA, USA, 21–23 June 1989.
8. Khalil, H.; Saberi, A. Adaptive stabilization of a class of nonlinear systems using high-gain feedback. *IEEE Trans. Autom. Contr.* **1987**, *32*, 1031–1035. [CrossRef]
9. Saberi, A.; Sannuti, P. Observer design for loop transfer recovery and for uncertain dynamical systems. *IEEE Trans. Autom. Contr.* **1990**, *35*, 878–897. [CrossRef]
10. Tornambe, A. Use of asymptotic observers having high-gains in the state and parameter estimation. In Proceedings of the 28th IEEE Conference on Decision and Control, Tampa, FL, USA, 13–15 December 1989; pp. 1792–1794.
11. Yin, Z.; Zhang, Y.; Du, C.; Liu, J.; Sun, X.; Zhong, Y. Research on Anti-Error Performance of Speed and Flux Estimation for Induction Motors Based on Robust Adaptive State Observer. *IEEE Trans. Ind. Electron.* **2016**, *63*, 3499–3510. [CrossRef]
12. Guo, B.-Z.; Zhao, Z.-L. On convergence of the nonlinear active disturbance rejection control for MIMO systems. *SIAM J. Control Optim.* **2013**, *51*, 1727–1757. [CrossRef]
13. Guo, B.-Z.; Zhao, Z.-L. On the convergence of an extended state observer for nonlinear systems with uncertainty. *Syst. Control Lett.* **2011**, *60*, 420–430. [CrossRef]
14. Zheng, Q.; Gao, L.; Gao, Z. On stability analysis of active disturbance rejection control for nonlinear time-varying plants with unknown dynamics. In Proceedings of the 2007 46th IEEE Conference on Decision and Control, New Orleans, LA, USA, 12–14 December 2007; pp. 12–14.
15. Chiasson, J. : *Modeling and High Performance Control of Electric Machines*; John Wiley and Sons: Hoboken, NJ, USA, 2005.
16. Abbas, H.A.; Belkheiri, M.; Zegnini, B. Feedback Linearization Control of an Induction Machine Augmented by Single Hidden Layer Neural Networks. *Int. J. Control* **2016**, *89*, 140–155. [CrossRef]
17. Zhang, Y.; Jiang, Z.; Yu, X. Indirect Field-Oriented Control of Induction Machines Based on Synergetic Control Theory. In Proceedings of the IEEE Power and Energy Society General Meeting—Conversion and Delivery of Electrical Energy in the 21st Century, Pittsburgh, PA, USA, 20–24 July 2008; pp. 1–7. [CrossRef]

engineering proceedings

MDPI

Proceeding Paper

Photovoltaic Panel Parameters Estimation Using an Opposition Based Initialization Particle Swarm Optimization †

Cilina Touabi *, Abderrahmane Ouadi and Hamid Bentarzi

Laboratory Signals and Systems, IGEE, University M'hamed Bougara, Boumerdes 35000, Algeria
* Correspondence: c.touabi@univ-boumerdes.dz

† Presented at the 2nd International Conference on Computational Engineering and Intelligent Systems, online, 18–20 November 2022.

Abstract: A photovoltaic (PV) cell is generally used as renewable energy source. For an accurate study of various PV applications, modeling this basic device in a PV generator is essential. However, the manufacturers do not usually provide the model parameters and their values vary over time due to PV degradation and the change in weather conditions. Thus, finding an optimal technique for estimating the appropriate parameters is crucial. This problem can be solved by metaheuristic optimization algorithms, namely particle swarm optimization (PSO). However, early convergence is the main defect of PSO. This work presents an enhancement in the optimization method (PSO) for identifying the optimal parameters of a PV generating unit. In this method, the identification of parameters of the single diode model is based on an opposition-based initialization particle swarm optimization technique. The optimization algorithm is implemented in MATLAB which gives good results.

Keywords: photovoltaic (PV) cell; opposition-based particle swarm optimization algorithm; one diode model parameters

1. Introduction

Solar Energy is considered as the most promising alternative to conventional energy sources. Its main application is the photovoltaic (PV) power generation that was predicted to be over 1000 TWh in 2021 [1]. PV systems convert solar energy into electrical energy. They can be installed easily, and they are noise-free. For an accurate study of various PV applications, modeling the basic device in a PV generator is essential. However, the manufacturers do not usually provide the model parameters and their values vary over time due to PV degradation [2] and the change in weather conditions. Thus, finding an optimal method for estimating appropriate parameters is critical.

The single diode model (ODM) is regarded as the most suitable model used to characterize the PV generator [3–7] compared to the two-diode model (DDM) and the three-diode model (TDM) as it has the least number of parameters and a good accuracy. The ODM has five electrical parameters that are: photocurrent (I_{ph}), reverse saturation current (Is), diode ideality factor (n), series resistance (Rs), and parallel resistance (R_{sh}).

Different techniques have been developed to identify ODM parameters, which can be classified in three categories [5,7]:

- Analytical methods [7],
- Numerical methods [3–6],
- Optimization methods based on artificial intelligence [8–11].

Although numerical methods are widely used in the literature due to their speed of calculation, simplicity, and accuracy, they cannot be used to solve PV model complex non-linear equations. This problem can be solved using optimization techniques based on artificial intelligence. Particle swarm optimization (PSO) is a popular metaheuristic

Citation: Touabi, C.; Ouadi, A.; Bentarzi, H. Photovoltaic Panel Parameters Estimation Using an Opposition Based Initialization Particle Swarm Optimization. *Eng. Proc.* **2023**, *29*, 16. https://doi.org/10.3390/engproc2023029016

Academic Editors: Abdelmadjid Recioui and Fatma Zohra Dekhandji

Published: 3 February 2023

optimization algorithm. However, the main defect of PSO that does not let it provide high-quality solutions in multimodal problems, such as PV panels parameters estimation, is its early convergence. This work presents an enhancement in the optimization method (PSO) for extracting the optimal parameters of a PV generating unit. The purpose of this work is to simulate the I-V and P-V characteristics based on the single diode model (ODM) using an opposition-based initialization particle swarm optimization algorithm that allows finding the optimal values of the needed parameters. The optimization algorithm is implemented in MATLAB for obtaining these model parameters and hence, I-V and P-V characteristics. The analysis is performed on various PV modules under different environmental conditions. The obtained results are compared and discussed to prove the efficiency and accuracy of the suggested optimization technique.

2. Photovoltaic Cell

A PV cell is an electronic device that permits the conversion of solar energy into electrical energy based on the photovoltaic effect, as shown in Figure 1. Solar cell produces electricity with poor voltage, which is approximately about 0.5 to 0.6 volts for the common single junction silicon PV cell. Thus, PV cells are coordinated in the form of modules or panels to produce electricity with high voltage and to provide adequate voltage and current for life applications.

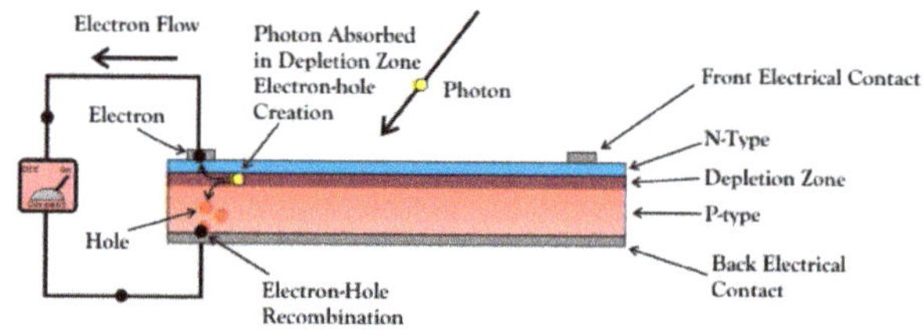

Figure 1. The photovoltaic effect in a PV cell.

2.1. Characteristics of the PV Cell

The electrical characteristics of a PV generator are mainly provided by the manufacturers under standard test conditions (STC) that are specified by the ambient temperature $T_{STC} = 25\ ^\circ C$, irradiation level $G_{STC} = 1000\ W/m^2$, and the air mass value AM = 1.5. However, in the working field, PV panels operate at varying temperatures and at lower insulation levels. In order to define the power output of the solar generator, it is essential to find the expected operating temperature of the PV panel. The nominal operating cell temperature (NOCT) is set as the temperature reached by open circuited cells in a PV panel under the conditions: ambient temperature $T_{ambient} = 20\ ^\circ C$, solar irradiance $G = 800\ w/m^2$, and wind speed = 1 m/s. Hence, the PV cell temperature can be calculated as follows:

$$T_{cell} = T_{ambient} + \left(\frac{NOCT - 20}{800}\right) * G \quad (1)$$

The typical I-V and P-V curves characterizing a photovoltaic cell are shown in Figure 2. The three significant points on the photovoltaic characteristics are short circuit current (I_{sc}), open circuit voltage (V_{oc}), and maximum power point (V_{mpp},I_{mpp}).

The maximum current I_{sc} in the photovoltaic cell is generated when a short circuit occurs between its terminals, while the maximum voltage V_{oc} can be measured when there is an open circuit.

The maximum power achieved from a photovoltaic cell occurs at a point on the bend in the I-V curve known as the maximum power point (MPP). The voltage and current at those points are designated as V_{mpp} and I_{mpp}.

Generally, manufacturers provide these parameters in the datasheet under STC. When the PV panel is connected to an external load, the actual point on the I-V curve at which the photovoltaic cell operates is determined based on the electrical characteristics of the load.

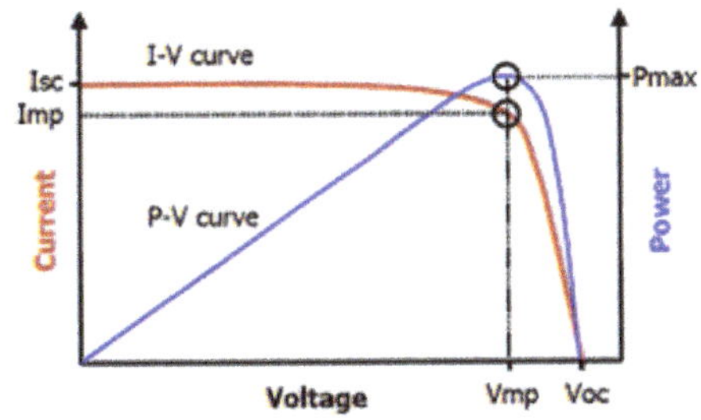

Figure 2. Typical I-V and P-V characteristics of PV cell.

2.2. One Diode Model of PV Cell

In order to analyze solar cells characteristics, this latter is modeled as electrical equivalent circuits using simulation software. Researchers have developed mathematical models to understand and study the effect of different weather conditions on photovoltaic electrical output. One of these models that is widely used is the lumped parameter model. This model is classified based on the number of diodes used and it has proven to be the most successful.

Although the model characteristics accuracy improves as the number of diodes increases, the model mathematical equation becomes more complex. In this work, the single diode model that is shown in Figure 3 is used for the identification of the PV generator parameters due to its simplicity compared to other lumped models and its good accuracy.

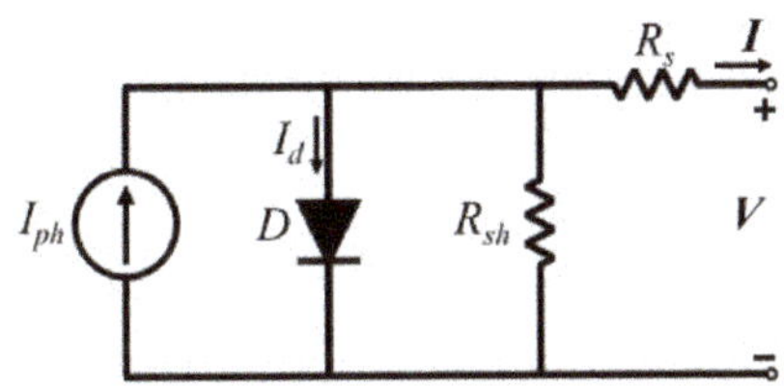

Figure 3. One diode model of PV cell.

The ODM governing equation is given as:

$$I = I_{ph} - I_s\left[e^{\left(\frac{V+I*R_s}{n*V_t}\right)} - 1\right] - \frac{V + I * R_s}{R_{sh}} \quad (2)$$

Such that, V_t is the thermal voltage.

The model's five parameters are:

I_{ph}: Photocurrent (A); I_s: Diode saturation current (A); n: Diode ideality factor; R_{sh}: Parallel resistance (Ω); R_s: Series resistance (Ω).

The single diode model considers various properties of the solar cell such as: the shunt resistance that considers the leakage current to the ground when the diode is in reverse biased, and the series resistance that takes into account the voltage drops and internal losses due to the flow of current. However, this model is still not the most accurate model as it has neglected the recombination effect of the diode.

3. Identification of PV Cell Parameters Using Optimization algorithms

3.1. Optimization Algorithms

It can be noticed that the I-V relationship of the one diode model, which is given in Equation (2) and hence the current voltage curve, is complex and highly nonlinear. Thus, it is difficult to solve it using analytical methods. Consequently, metaheuristic optimization methods based on artificial intelligence have been developed by scientists for solving these kinds of equations and determine the needed parameters. These techniques are developed to find a good solution among a large set of feasible solutions with less computational effort than other optimization techniques.

3.2. Opposition Based Initialization Particle Swarm Optimization Technique

Particle Swarm Optimization is a population-based metaheuristic global optimization method inspired by the motion of schooling fish and bird flocks. The PSO algorithm examines the space of an objective function by adjusting the trajectories of individual agents, named particles. A population of these particles flies through the search space such that each particle i is attracted toward the position of the current global best g^* and its own best location x^i in history, simultaneously, it has a tendency to move randomly. Initially, the particles are placed randomly in the search space. The objective function is evaluated for all the particles. When a particle i finds a position that is better than any previously found locations, it updates it as the new current best location x^i by updating the velocity, first depending on movement inertia, self-cognition, and social interaction using Equation (3), then updating its position through Equation (4) at each iteration. Thus, all n particles have a current best position at any time t during iterations. The purpose is to find the global best among all the current best solutions until the objective no longer improves or after a specified number of iterations. The motion of particles is schematically represented in Figure 4. Where $g^* = \{(x^i)\}$ for (i = 1, 2, … n) is the current global best and x_i^* is the current best for particle i [8,12].

$$v_{n+1}^{i} = w\, v_{n}^{i} + c_1 r_1 \left[p_{n}^{i} - x_{n}^{i}\right] + c_2 r_2 \left[p_{n}^{g} - x_{n}^{i}\right] \quad (3)$$

$$x_{n+1}^{i} = x_{n}^{i} + v_{n+1}^{i} \quad (4)$$

Such that:
x^i is the position of the ith particle in the search space;
v^i is the velocity of the ith particle;
w is particle inertia;
c_1 is the cognitive acceleration constant;
c_2 is the social acceleration constant;
p^i is the particle's best-known position;
p^g is the global best position;
r_1, r_2 are random numbers that vary between 0 and 1.

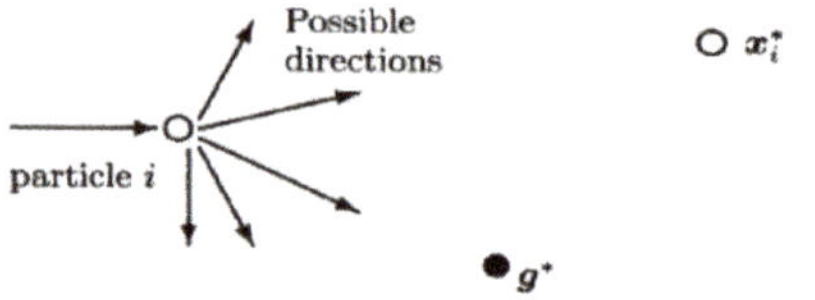

Figure 4. Schematic representation of the motion of a particle in PSO.

The starting points in PSO are randomly given. If these latter are close to the optimal point, convergence speed would be faster. Thus, to get better results, a better and careful initialization based on priori information is needed. The proposed method for an improved PSO algorithm is an opposition-based initialization of the swarm. This approach consists of initializing the PSO population and its opposite population, as shown in Figure 5. The fitness function is evaluated for both populations and only the fitter particles are selected to form a new population for the PSO.

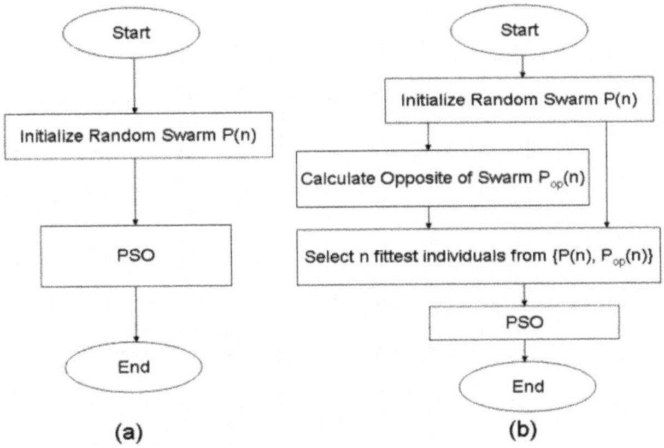

Figure 5. PSO with (**a**) Random population initialization and (**b**) Opposition based population initialization.

The concept used is described below.
Particle: a swarm particle p^i is defined as:
$i \in [a, b]$ such that, $i = 1, 2, \ldots, D$ and $a, b \in R$;
D represents dimensions, and R represents real numbers.
Opposite particle: every particle p^i has a unique opposite $p^i{}_{op}$ defined as:

$$p^i{}_{op} = a + b - p^i \text{ such that, } i = 1, 2, \ldots D \text{ and } a, b \in R \tag{5}$$

3.3. ODM Parameters Extraction Using an IOB-PSO

The following objective function is used,

$$\begin{cases} F(X) = I - I_{ph} + I_s\left[e^{\left(\frac{V+I*R_s}{n*N_s*V_t}\right)} - 1\right] + \frac{V+I*R_s}{R_{sh}} \\ X = \{I_{ph}, I_s, R_s, n, R_{sh}\} \end{cases} \tag{6}$$

The fitness function used to quantify the error between the simulated and measured data is the root mean square error (RMSE),

$$\text{Fitness} = \sqrt{\frac{1}{N}\sum_{1}^{N} F(X)^2} \tag{7}$$

The pseudo code of the IBPSO method is given in Algorithm 1 and its corresponding flowchart is shown in Figure 6.

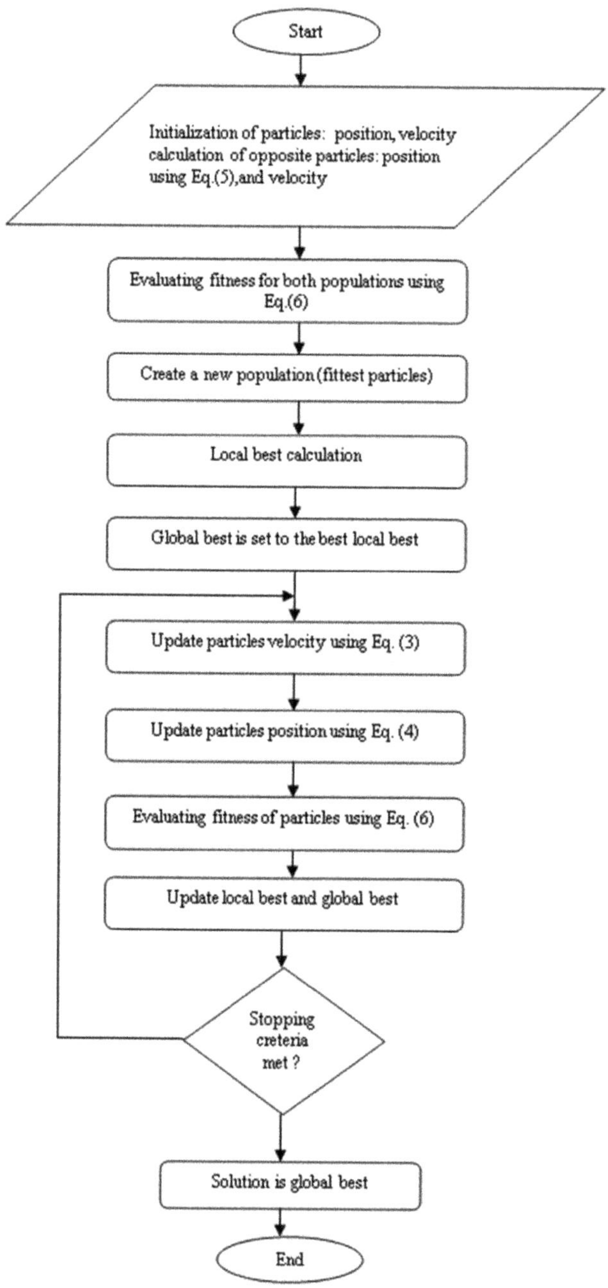

Figure 6. The flowchart of the IOBPSO algorithm.

Algorithm 1 IOBPSO

```
Input: T, Ns, V_Data , I_Data
Output: global best
for each particle
        Initialize particle position
        Initialize particle velocity
        Calculate cost value of particles using Equation (6)
        Initialize opposite particle position using Equation(5)
        Initialize opposite particle velocity
        Calculate cost value of opposite particles using Equation (6)
    If opposite particle cost < particle cost
    Update particle
    end if
    Update particle best
    If particle best cost < global best cost
    Update global best
    end if
end for
for (t =1: Max number of iterations)
    for each particle
            Update velocity using Equation (3)
    Update position using Equation(4)
    Calculate cost value using Equation (6)
    If particle cost < particle best cost
        Update particle best
        If particle best cost < global best cost
            Update global best
        end if
    end if
    end for
end for
return global best
end procedure
```

4. Test Results and Discussion

The proposed algorithm IOB-PSO is used to identify the parameters of the single diode model. The algorithm is tested for two different PV modules. The obtained results are compared with other methods results to prove its effectiveness. It is developed in MATLAB R2016a and executed under Windows 10 64-bit OS, on a PC with Intel® Core™ i5-2450M CPU processor @ 2.50GHz, 4GB RAM.

Table 1 presents the search ranges used for the optimization of the model parameters. Table 2 presents the IOB-PSO parameters.

Table 1. ODM parameter search ranges.

Parameters	Search Range
I_{ph}	$[0.95 \times I_{sc}, 1.05 \times I_{sc}]$
I_s	$[1\ \mu A, 5\ \mu A]$
n	[1, 2]
R_{sh}	$[\frac{V_{mpp}}{I_{sc} - I_{mpp}}, 1500\ \Omega]$
R_s	$[0, \frac{V_{mpp} - V_{oc}}{I_{mpp}}]$

Table 2. IOB-PSO parameters.

Parameters	Value
Cognitive factor c_1	1.5
Social factor c_2	2.0
Inertia weight w	[0.2, 0.9]
Random values r_1, r_2	[0, 1]
Number of particles	30
Maximum iteration	1000

The IOB-PSO algorithm is used to estimate parameters for the various PV modules. The electrical specifications of the utilized modules are described in Table 3.

Table 3. Electrical specifications of the PV modules.

Module	Type	N_s	Temperature [°C]	Irradiance [w/m^2]
STM6-40/36	Mono-crystalline	36	51	NA
Photowatt-PWP201	Poly-crystalline	36	45	1000

NA: not available.

The IOB-PSO is tested for the mono-crystalline STM6-40/36 module using I-V experimental data [13] measured at T = 51 °C; and for the Photowatt-PWP201 poly-crystalline module with I-V data measured under a temperature of T = 45 °C [13]. The obtained results are presented and compared with other previously published methods in Tables 4 and 5, respectively.

Table 4. STM6-40/36 extracted parameters achieved by various methods.

Meth.	Parameters					Error
	I_{ph}[A]	I_s[μA]	n	R_s[Ω]	R_{sh}[Ω]	RMSE
ABC [14]	1.50	1.664	1.487	4.99	15.21	1.838×10^{-3}
CIABC [14]	1.664	1.676	1.498	4.40	15.62	1.819×10^{-3}
CSA [13]	1.664	2.000	1.534	2.91	15.841	1.794×10^{-3}
ImCSA [13]	1.664	2.000	1.534	2.92	15.841	1.795×10^{-3}
IOB-PSO *	1.663	2.88	1.57	0.0015	598.74	1.772×10^{-3}

* Proposed method.

Table 5. Photowatt-PWP201 extracted parameters achieved by different methods.

Meth.	Parameters					Error
	I_{ph}[A]	I_s[μA]	n	R_s[Ω]	R_{sh}[Ω]	RMSE
CPSO [13]	1.0286	8.3010	1.451194	1.0755	1850.1	3.5×10^{-3}
PS [13]	1.0313	3.1756	1.341358	1.2053	714.2857	1.18×10^{-2}
SA [13]	1.0331	3.6642	1.356142	1.1989	833.3333	2.7×10^{-3}
CARO [15]	1.03185	3.28401	1.35453	1.20556	841.3213	2.427×10^{-3}
IOB-PSO *	1.030	3.495668	1.35	1.200877	986.306335	2.4251×10^{-3}

* Proposed method.

Using the identified parameters, the PV characteristics of the two modules were constructed and then compared with the experimental curves in Figures 7 and 8.

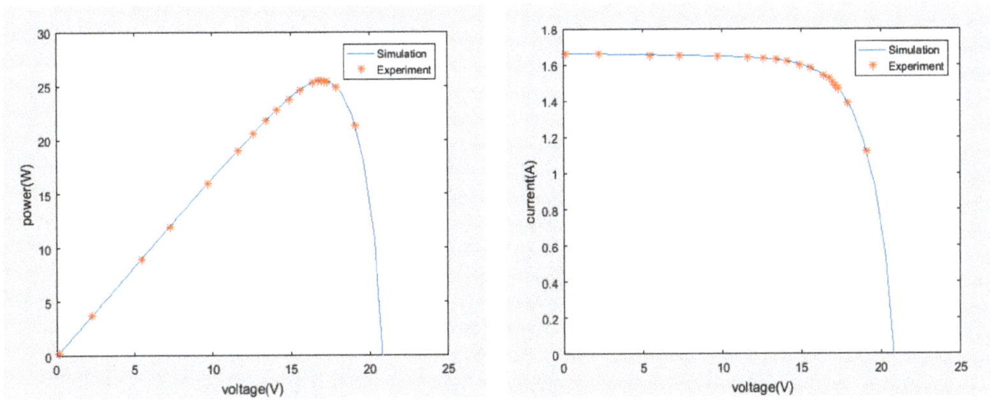

Figure 7. I-V and P-V experimental data and simulated data for the STM6−40/36 module.

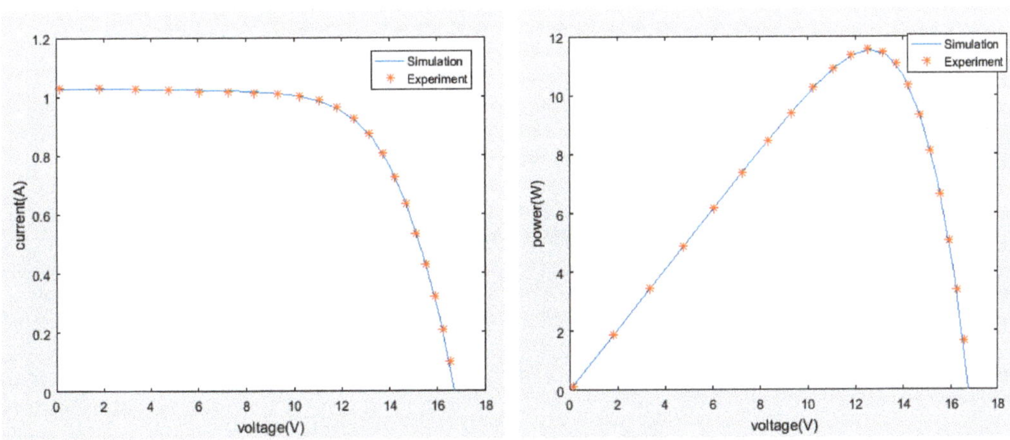

Figure 8. I-V and P-V experimental data and simulated data for the Photowatt-PWP201 module.

It can be noted that the RMSE error of the proposed IOB-PSO is the smallest and hence, the obtained results using the proposed algorithm are the best. Therefore, the efficiency of this method for identifying the model parameters is proven. Figures 7 and 8 show the comparison between I-V and P-V experimental and simulated data of the STM6-40/36 and the Photowatt-PWP201 modules. The simulated curves highly match the measured ones, which proves the reliability of the IOB-PSO algorithm.

Figure 9 presents the convergence curve of the IOB-PSO method for these modules; it can be noticed that the convergence is very fast. The algorithm reached the optimal solution within 207 iterations in an execution time of 19.0017 s for the monocrystalline module STM6-40/36, while for the Photowatt-PWP201 module, it needed 613 iterations (27.585 s). This proves that the proposed IOB-PSO method has a very fast convergence speed.

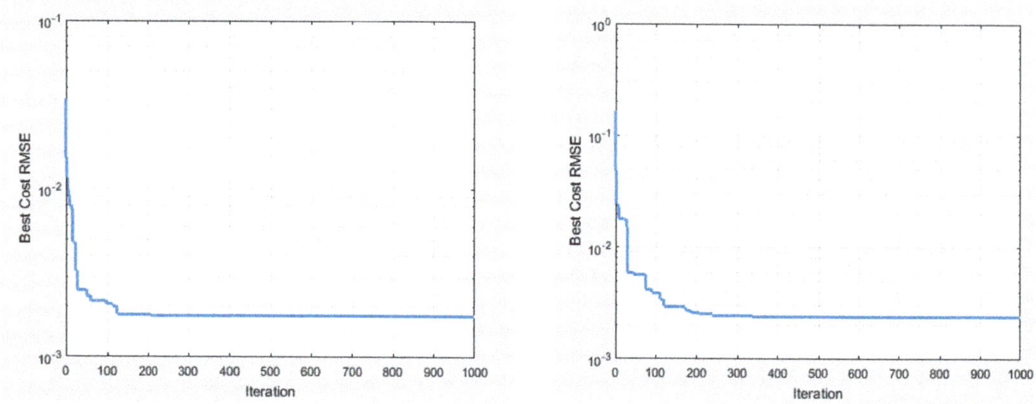

(a) (b)

Figure 9. Convergence curve of the IOB-PSO for the (**a**) STM6−40/36 module (**b**) Photowatt-PWP201 module.

5. Conclusions

In this work, an efficient method for obtaining the I-V and P-V characteristics of a PV module based on the one diode model has been proposed, due to the non-availability of all needed parameters in the datasheet. The optimization algorithm was chosen for the identification of the ODM model parameters. The proposed method is an improved opposition-based particle swarm optimization; this algorithm was tested using the experimental I-V data that were acquired at different working conditions for various PV modules. The obtained results were compared with other methods outcomes provided in the references; this algorithm has showed a satisfying estimation of the five parameters extracted with minimum errors. Furthermore, the PV characteristics were plotted using the extracted parameters and compared with the experimental data. The simulated curves are found to be well-suited with the corresponding measured data, thus, the performance of the IOB-PSO algorithm proved to be good. Adding to the fact that this proposed algorithm has provided optimal results with an acceptable accuracy, the time taken for the IOB-PSO execution is less than 30 s.

This approach is found to be useful for designers since it is simple, fast, and provided accurate results.

Author Contributions: Conceptualization, C.T., A.O. and H.B.; methodology, C.T., A.O. and H.B.; software, C.T.; validation, C.T., A.O. and H.B.; formal analysis, C.T., A.O. and H.B.; investigation, C.T., A.O. and H.B.; resources, C.T.; data curation, C.T.; writing—original draft preparation, H.B.; writing—review and editing, C.T. and H.B.; visualization, H.B.; supervision, A.O. and H.B.; project administration, A.O. and H.B.; funding acquisition, H.B. All authors have read and agreed to the published version of the manuscript.

Funding: This research received no external funding.

Institutional Review Board Statement: Not applicable.

Informed Consent Statement: Not applicable.

Data Availability Statement: The data presented in this study are available on request from the corresponding author. The data are not publicly available due to the data may involve confidential information of our research group.

Conflicts of Interest: The authors declare no conflict of interest.

References

1. Solar PV—Analysis. IEA Tracking Report—September 2022. Available online: https://www.iea.org/reports/solar-pv (accessed on 4 January 2023).
2. Xiong, G.; Zhang, J.; Shi, D.; He, Y. Parameter extraction of solar photovoltaic models using an improved whale optimization algorithm. *Energy Convers. Manag.* **2018**, *174*, 388–405. [CrossRef]
3. Stornelli, V.; Muttillo, M.; De Rubeis, T.; Nardi, I. A new simplified five parameter estimation method for single-diode model of photovoltaic panels. *Energies* **2019**, *12*, 4271. [CrossRef]
4. Silva, E.A.; Bradaschia, F.; Cavalcanti, M.C.; Nascimento, A.J. Parameter estimation method to improve the accuracy of photovoltaic electrical model. *IEEE J. Photovolt.* **2016**, *6*, 278–285. [CrossRef]
5. Wang, G.; Zhao, K.; Shi, J.; Chen, W.; Zhang, H.; Yang, X.; Zhao, Y. An iterative approach for modeling photovoltaic modules without implicit equations. *Appl. Energy* **2017**, *202*, 189–198. [CrossRef]
6. Orioli, A. An accurate one-diode model suited to represent the current-voltage characteristics of crystalline and thin-film photovoltaic modules. *Renew. Energy* **2020**, *145*, 725–743. [CrossRef]
7. Batzelis, E. Non-iterative methods for the extraction of the single-diode model parameters of photovoltaic modules: A review and comparative assessment. *Energies* **2019**, *12*, 358. [CrossRef]
8. Wei, T.; Yu, F.; Huang, G.; Xu, C. A particle-swarm-optimization based parameter extraction routine for three-diode lumped parameter model of organic solar cells. *IEEE Electron Device Lett.* **2019**, *40*, 1511–1514. [CrossRef]
9. Zhang, Y.; Jin, Z.; Zhao, X.; Yang, Q. Backtracking search algorithm with hevy flight for estimating parameters of photovoltaic models. *Energy Convers. Manag.* **2020**, *208*, 112615. [CrossRef]
10. Jiao, S.; Chong, G.; Huang, C.; Hu, H.; Wang, M.; Heidari, A.A.; Chen, H.; Zhao, X. Orthogonally adapted harris hawks optimization for parameter estimation of photovoltaic models. *Energy* **2020**, *203*, 117804. [CrossRef]
11. Naeijian, M.; Rahimnejad, A.; Ebrahimi, S.M.; Pourmousa, N.; Gadsden, S.A. Parameter estimation of PV solar cells and modules using Whippy Harris Hawks Optimization Algorithm. *Energy Rep.* **2021**, *7*, 4047–4063. [CrossRef]
12. Ahcene, F.; Bentarzi, H.; Ouadi, A. Automatic Voltage Regulator Design Using Particle Swarm Optimization Technique. In Proceedings of the 2020 International Conference on Electrical Engineering (ICEE), Istanbul, Turkey, 25–27 September 2020.
13. Kang, T.; Yao, J.; Jin, M.; Yang, S.; Duong, T. A Novel Improved Cuckoo Search Algorithm for Parameter Estimation of Photovoltaic (PV) Models. *Energies* **2018**, *11*, 1060. [CrossRef]
14. Oliva, D.; Ewees, A.A.; Aziz, M.A.E.; Hassanien, A.E.; Cisneros, M.P. A chaotic improved artificial bee colony for parameter estimation of photovoltaic cells. *Energies* **2017**, *10*, 865. [CrossRef]
15. Yuan, X.; He, Y.; Liu, L. Parameter extraction of solar cell models using chaotic asexual reproduction optimization. *Neural Comput. Appl.* **2015**, *26*, 1227–1239. [CrossRef]

engineering proceedings

Proceeding Paper

APF Applied on PV Conversion Chain Network Using FLC †

Bourourou Fares *, Tadjer Sid Ahmed and Habi Idir

LREEI, M'hamed Bougara of Boumerdes University, Boumerdes 35000, Algeria
* Correspondence: f.bourourou@univ-boumerdes.dz

† Presented at the 2nd International Conference on Computational Engineering and Intelligent Systems, Online, 18–20 November 2022.

Abstract: This paper focuses on regulation of the parallel active power filter (APF) Dc Voltage bus by judicious choice of rule bases and intervals for each selected fuzzy variable of suitable fuzzy logic controller. In addition, an algorithm describes the main steps for designing an FLC that has any number of rules with direct application to the APF capacitor voltage regulation. Where their simulation, by MATLAB, applied to PV conversion chain network will be represented in the booths cases, constant and variable non-linear loads after modeling, to show the effectiveness of this kind of regulators on electrical power quality and improve the reliability of the APF on PV system. The delivered voltage of PV plant has been regulated and controlled with MPPT using P&O technique and FLC regulator after modeling of each part of the conversion chain. PV plant supplies a nonlinear load from the rectifier installed on the output of the conversion chain via a controlled power inverter. A 3 × 3 rules fuzzy regulator is implanted in the control part of the APF to examine the influence of the FLC on the produced electrical power quality. Simulation results are represented and analyzed.

Keywords: APF; PV; FLC; renewable energy; power quality

1. Introduction

PV plant conversion chain description and the MPPT technique based on P&O under irradiation and temperature variation influence the generated amplitude of output voltage [1–3] and the detailed modeling behavioral Matlab simulation. For obtaining good performance and efficiency energy, it is proposed to host a diverse suitable controls strategy to replace the power electronic interface to achieve the needed performance results for the system [3].

Citation: Fares, B.; Ahmed, T.S.; Idir, H. APF Applied on PV Conversion Chain Network Using FLC. *Eng. Proc.* **2023**, *29*, 17. https://doi.org/10.3390/engproc2023029017

Academic Editors: Abdelmadjid Recioui, Hamid Bentarzi and Fatma Zohra Dekhandji

Published: 19 April 2023

2. Description and Modeling of PV Conversion Chain

2.1. PV System Description

The global scheme present in Figure 1 shows the main parts dedicated to converting sun power to electrical energy. The use of PV allows the conversion to be achieved, and the system will be modeled.

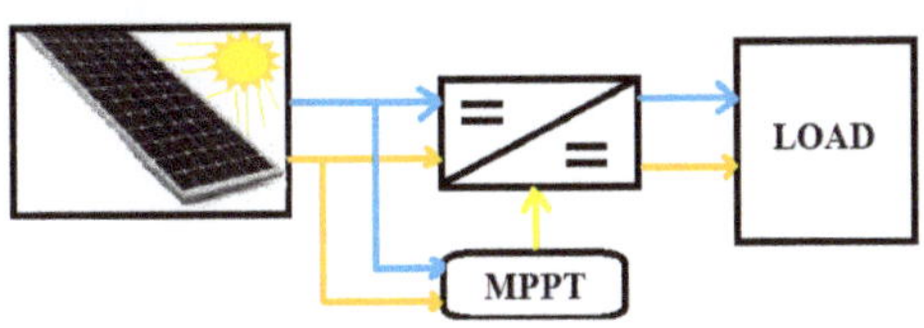

Figure 1. PV station power conversion basic part.

2.2. PV System Modeling

Solar cells' modeling is essential to the study of photovoltaic plant generators. Generally, it is represented by an equivalent circuit [4] shown in Figure 2 below.

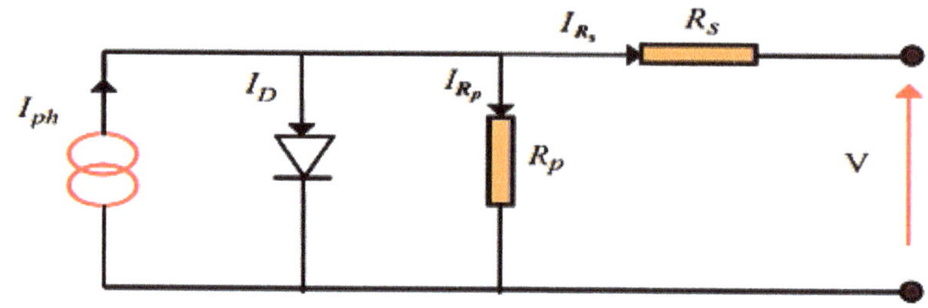

Figure 2. One-diode equivalent circuit of the solar cell.

From Figure 2, for equivalent circuit, we can obtain

$$I_{ph} = I_D + I_{Rp} + I. \quad (1)$$

The resistor R_p current is obtained by

$$I_{Rp} = \frac{V + I_{Rs}}{R_p}, \quad (2)$$

and diode current is obtained by

$$I_D = I_s\left[e^{\frac{(V+I_{Rs})}{nV_t}} - 1\right], \quad (3)$$

where I_s is the diode saturation current obtained by

$$I_s = K_1 T^3 e^{\frac{E_g}{KT}}, \quad (4)$$

where
$V_t = KT/q$: Thermal stress at temperature T;
q: Electron charge $\left(1\,.602 \times 10^{-19}\ \mathrm{C}\right)$;
K: Boltzmann constant $\left(1.381 \times 10^{-23}\ \mathrm{J/k}\right)$;
K_1: Constant $\left(1.2\ \mathrm{A/cm^2K^3}\right)$;
n: Junction non-ideality factor;
T: Effective cell temperature in Kelvin;
E_g: Gap energy (for crystalline silicon is equal to 1.12 eV).

Therefore, the expression of the characteristic I (V) is

$$I = I_{ph} - I_s\left[e^{\frac{(V+I_{Rs})}{nV_t}} - 1\right] - \frac{V + I_{Rs}}{R_p}. \quad (5)$$

The application of the Newton method makes it possible to calculate the value of the current I for each iteration with

$$I_{n+1} = I_n - \frac{I_{cc} - I_n - I_s[e^{\frac{(V+R_S.I)}{nV_T}} - 1]}{-1 - I_s\left(\frac{R_s}{nV_T}\right).[e^{\frac{(V+I_n.R_S)}{nV_T}}]}. \quad (6)$$

Then, the new value of I_{cc}, short circuit courant, corresponds to an irradiation G, and a given temperature T is calculated according to the following equation:

$$I_{cc}(G,T) = I_{ccr}\frac{G}{1000}[1 + a\left(T - T_{ref}\right)]. \quad (7)$$

Diode saturation current depends on the temperature. Its value for a given temperature T can be calculated by

$$I_s(T) = I_{sr}\left(T_{ref}\right)\left(\frac{T}{T_{ref}}\right)^{\frac{3}{n}} e^{\left(\frac{-q.E_g}{nK}\right)\left(\frac{1}{T}-\frac{1}{T_{ref}}\right)}. \tag{8}$$

Influence of solar radiation and temperature on PV power is represented by Equation (9):

$$P_{PV} = \left(I_{sct} - N_P I_{0S}\left(e^{\frac{\frac{V_{PV}}{N_S}+\frac{R_S I_{PV}}{N_P}}{V_T}} - 1\right) - \frac{V_{PV}}{R_p} - \frac{R_S I_{PV}}{R_p}\right).V_{PV}. \tag{9}$$

A *PV* energy conversion system must have other components ensuring the operation of the system in a more reliable and optimum mode for power and efficiency indices, such as the use of DC-DC converters controlled by different control techniques [3–5], to have the desired output power able to supply industrial equipment.

2.3. PV System MPPT Dependency on T & Ir

The power delivered by a PV depends on the ambient temperature, the wind speed, the mounting of the module (integrated in the roof or ventilated), and all these parameters change according to the chosen site for module installation. In addition, the coefficients linked to the temperature differ according to the materials used for the manufacture of the module [4].

Temperature is a very important parameter in the behavior of PV cells [4]. Figure 3 describes the behavior of the module under a fixed illumination of 1000 W/m^2, and at temperatures between 15 °C and 40 °C. We notice that the current increases with the temperature; on the other hand, the open circuit voltage decreases. This leads to a decrease in the maximum power available.

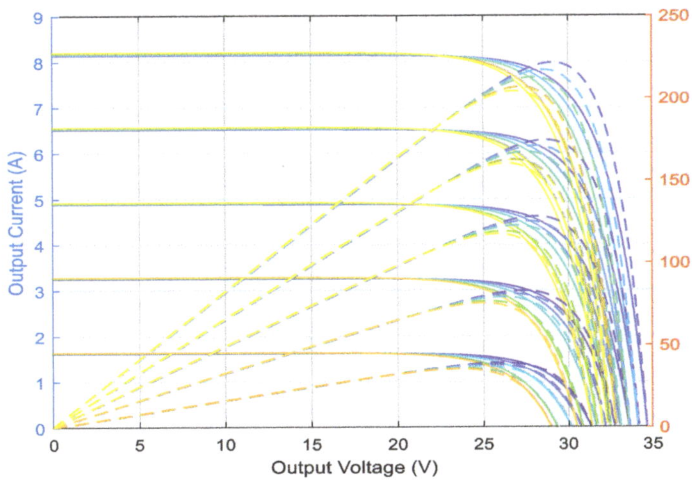

Figure 3. MPPT function of irradiation and temperature.

3. PV Power Quality Improvement Using APF and Simulation

The simulation block diagram presented in Figure 4 for an autonomous PV system based on polycrystalline solar panels under standard temperature and irradiation conditions (25 °C, 1000 W/m^2) which supplies a load via a DC-DC-AC converter is used to supply the industrial plant with integration of an APF on the output of the PV inverter to minimize the harmonic presence and assure the power quality in the acceptable range. A fuzzy logic controller is used to ensure the DC voltage control of the parallel APF which is based on rules shown in the figure below.

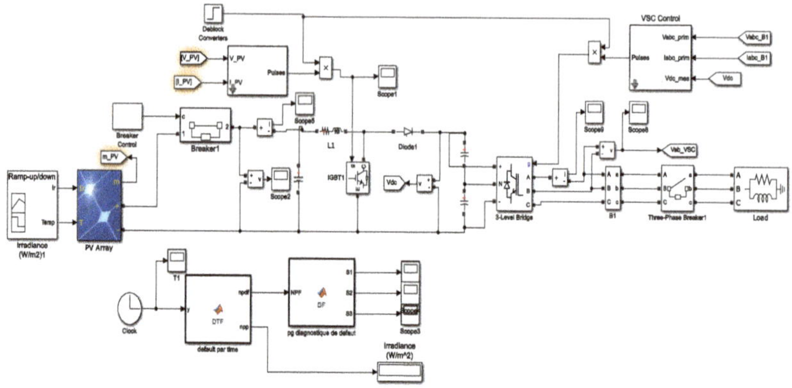

Figure 4. PV plant simulation bloc.

Where the member sheep functions of error and the error variation are chow on Figure 5 and the output control cis represented by Figure 6 below

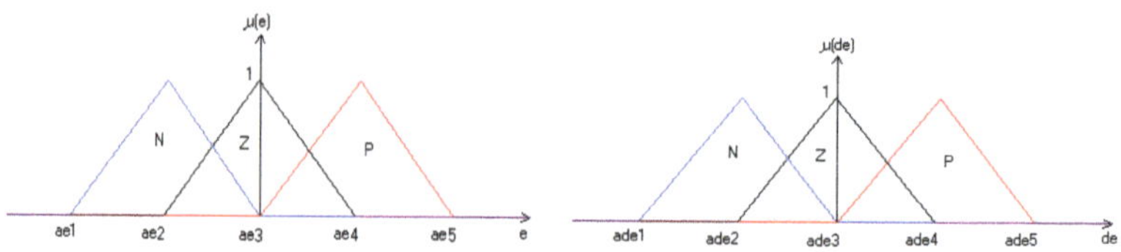

Figure 5. Error and error variation membership functions.

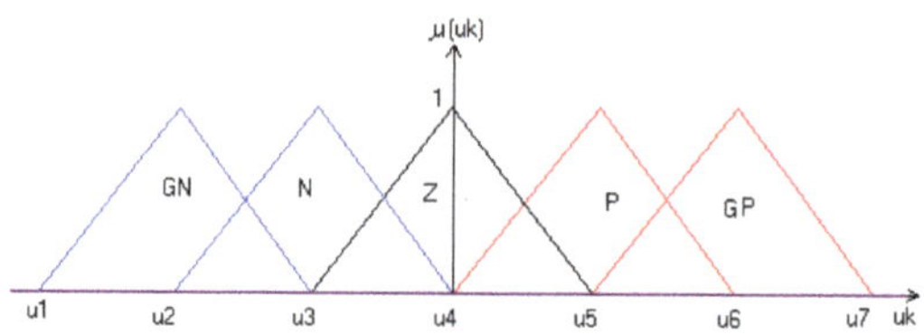

Figure 6. Command membership functions.

$$\mu_z[e(t)] = \begin{cases} 0 \text{ if } e < ae2 \\ (e - ae2)/(ae3 - ae2) \text{ if } ae2 < e < ae3 \\ (ae4 - e)/(ae4 - ae3) \text{ if } ae3 < e < ae4 \\ 0 \text{ if } e > ae4 \end{cases}, \tag{10}$$

where the inference table is given by Table 1.

Table 1. FLC inference table.

Δe/e	N	Z	P
N	GN	Z	Z
Z	N	Z	P
P	Z	Z	GP

The aggregation rules are the following:

if (e < 0 & Δe(t) < 0) ==> Ue is GN;
if (e > 0 et Δe(t) > 0) ==> U is GP;
if (e < 0 & Δe(t) > 0) **OR** (e = 0 & Δe(t) = 0) **OR** (e = 0 & Δe(t) > 0) **OR** (e = 0 & Δe(t) < 0);
OR (e > 0 & Δe(t) < 0) ====> U is Z ;
if (e < 0 & Δe(t) = 0) ==> U is N ;
if (e > 0 & Δe(t) = 0) ==> U is P.

The obtained results represented on Figures 7–9 show the effect of the FLC on power quality improvement and the APF Dc voltage regulation with a response time less than 0.01 s.

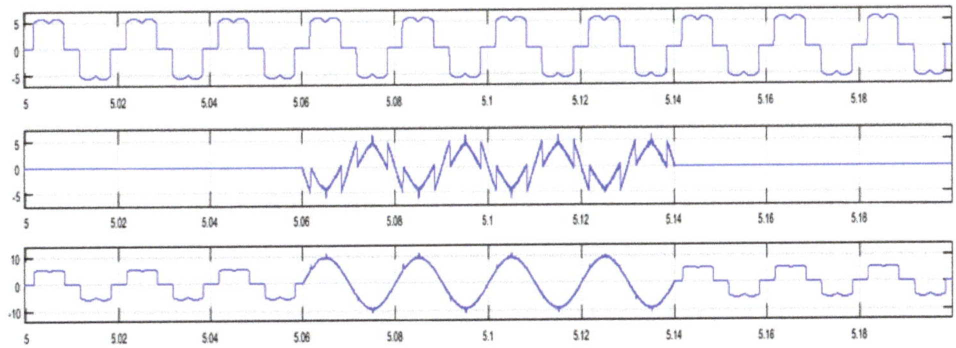

Figure 7. Load, APF and Supply current with and without APF connection.

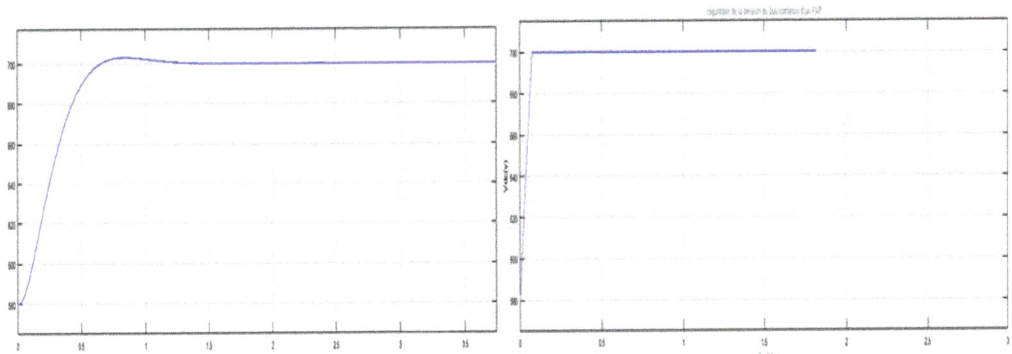

Figure 8. APF V_{DC} voltage regulation with PI then FLC controller.

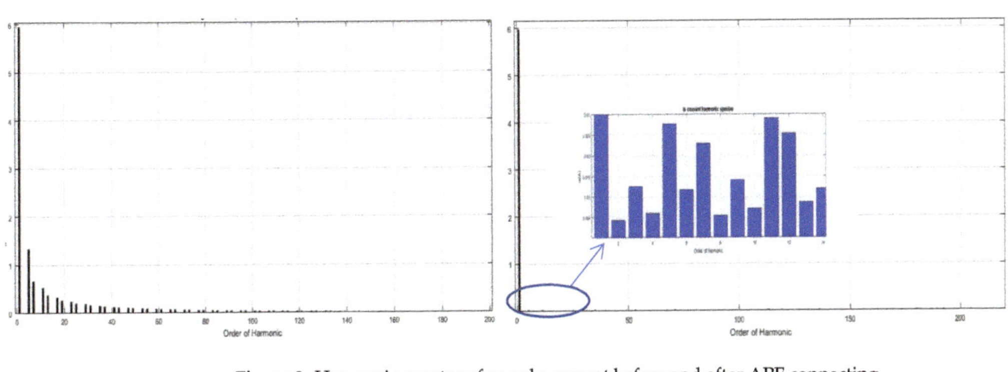

Figure 9. Harmonic specter of supply current before and after APF connecting.

Also static error obtained is near to zero in the two cases PI and FLC controllers results.

Where the harmonic specter represented on Figure 9 show the values of each harmonic before and after APF integration on Pv installation.

4. Conclusions

Nonlinear loads supplied by PV station harmonic have been described by the use of an APF controlled by PI controller and FLC controller. The APF installation on the network connected with the PV conversion chain decreased the THD of the network current from THD > 23% to less than THD < 5% with the use of a fuzzy logic controller.

We hope to apply more intelligent techniques to the studied system to obtain more suitable results for industrial application.

Author Contributions: Conceptualization, B.F.; methodology, B.F.; software, B.F.; validation, B.F.; formal analysis, B.F.; investigation, B.F.; resources, B.F.; data curation, B.F.; writing—original draft preparation, B.F.; writing—review and editing, B.F., T.S.A. and H.I.; visualization, B.F.; supervision, B.F., T.S.A. and H.I.; project administration, T.S.A. and H.I. All authors have read and agreed to the published version of the manuscript.

Funding: This research received no external funding for the university M'hamed Bougara of Boumerdes.

Institutional Review Board Statement: Validation by CS of LREE laboratory.

Informed Consent Statement: This study don't involving humans.

Data Availability Statement: The work has been elaborated in the university of M'hamed Bougara of Boumerdes, faculty of hydrocarbons and chemistry, LREE laboratory directed by Habi Idir.

Conflicts of Interest: The authors declare no conflict of interest.

References

1. Benchouia, M.; Ghadbane, I.; Golea, A.; Srairi, K.; Benbouzid, M. Implementation of adaptive fuzzy logic and PI controllers to regulate the DC bus voltage of shunt active power filter. *Appl. Soft Comput.* **2015**, *28*, 125–131. [CrossRef]
2. Zelloumaa, L.; Rabhib, B.; Saadc, S.; Benaissad, A.; Benkhorise, M.F. Fuzzy logic controller of five levels active power filter. *Energy Procedia* **2015**, *74*, 1015–1025. [CrossRef]
3. Li, H.-X.; Gatland, H.B. New methodology for designing a fuzzy logic controller. *IEEE Trans. Syst. Man Cybern.* **1995**, *25*, 3.
4. Hassan, A.A.; Fahmy, F.H.; El-Shafy, A.; Nafeh, A.; Hosam, K.M. Youssef, Control of Three-Phase Inverters for Smart Grid Integration of Photovoltaic Systems. *J. Electr. Syst.* **2022**, *18*, 109–131.
5. Kamel, K.; Laid, Z.; Kouzou, A.; Hafaifa, A.; Khiter, A. Comparison of Control Strategies for Shunt Active Power Filters in Three-Phase Three-Wire Systems. In Proceedings of the 3rd International Conference on Power Electronics and their Applications ICPEA 2017, Djelfa, Algeria, 16–17 September 2017.

MDPI
St. Alban-Anlage 66
4052 Basel
Switzerland
Tel. +41 61 683 77 34
Fax +41 61 302 89 18
www.mdpi.com

Engineering Proceedings Editorial Office
E-mail: engproc@mdpi.com
www.mdpi.com/journal/engproc

www.ingramcontent.com/pod-product-compliance
Lightning Source LLC
Chambersburg PA
CBHW061627210926
51850CB00009B/2465
9783036581125